Earthmasters

THE DAWN OF THE AGE OF
CLIMATE ENGINEERING

Clive Hamilton

YALE UNIVERSITY PRESS
NEW HAVEN AND LONDON

Copyright © 2013 Clive Hamilton

For information about this and other Yale University Press publications, please contact:
U.S. Office: sales.press@yale.edu   www.yalebooks.com
Europe Office: sales @yaleup.co.uk   www.yalebooks.co.uk

Set in Arno Pro by IDSUK (DataConnection) Ltd
Printed in Great Britain by TJ International Ltd, Padstow, Cornwall

Library of Congress Cataloging-in-Publication Data

Hamilton, Clive.
  Earthmasters : the dawn of the age of climate engineering/Clive Hamilton.
    pages cm
  Includes bibliographical references.
  ISBN 978-0-300-18667-3 (hardbound : alk. paper)
1. Weather control. 2. Climate change mitigation. 3. Environmental geotechnology. I. Title.
  QC928.H36 2013
  551.68—dc23
2012042932

A catalogue record for this book is available from the British Library.

9 8 7 6 5 4 3 2 1

The time is coming when the struggle for dominion over the earth will be carried on. It will be carried on in the name of fundamental philosophical doctrines.

Friedrich Nietzsche, 1882

Contents

Figures and Tables

Figures

Tables

Preface

For sheer audacity, no plan by humans exceeds the one now being hatched to take control of the Earth's climate. Yet it is audacity born of failure, a collective refusal to do what must be done to protect the Earth and ourselves from a future that promises to be nasty, brutish and hot. In my 2010 book *Requiem for a Species: Why We Resist the Truth about Climate Change*, I relied on the best science to set out that expected future. Even under optimistic assumptions about when the nations of the world will wake up to the danger and how quickly they will then respond, dramatic and long-lasting changes in the Earth's climate now seem unavoidable.

The task of *Requiem* was to explain why we – 'the rational animal' – have been unable to hear the insistent warnings of science academies around the world. It is into the yawning gap between the urgent response scientists say is needed and the timid measures governments are willing to take that geoengineering has stepped. Its promise seems irresistible – it is cheap, effective and free of the unpalatable side-effects of carbon abatement, such as the wrath of fossil fuel corporations and the resentment of voters willing to make only symbolic changes to their ways.

Superficially, climate engineering (here synonymous with geoengineering) appears to be another application of human technological ingenuity; yet, I will suggest, it actually represents a profound change in the relationship of Homo sapiens to the Earth. In the twenty-first century the fate of nature has come to depend on the 'goodwill' of humans, and to the extent that humans are part of nature the Earth system itself has acquired a moral force. So geoengineering is not just a technological problem, nor even an ethical one as usually understood; it goes to the heart of what it means for one species to hold the future of a planet in its hands. While most of this book is devoted to the who, what, why, where and when of geoengineering, I hope it will provoke the reader to reflect on a deeper question: What have we become?

Acknowledgements

When lay people write about science they had better get it right, especially in the case of climate science. So I am particularly grateful to the following scientists who have commented on all or parts of chapters 2 and 3: Graeme Pearman, Roger Gifford, Andrew Glikson, Tim Kruger, Phil Renforth, David Mitchell, Mike Raupach and Stephen Salter.

I am doubly grateful to Andrew Glikson for his very helpful commentary on other parts of the manuscript. Geoff Symons also provided useful suggestions. A reader invited by the publisher to comment provided sound advice on how to improve the book.

Will Steffen and Paul Crutzen provided valuable advice on the material in the last chapter, and Mark Lawrence helped me get the history right in the first chapter. Hugh Gusterson kindly commented on the Livermore story. Simon Terry and Pat Mooney provided important information on international negotiations. Clare Heyward challenged me to clarify and defend my views on ethics and Andrew Parker provided valuable background along the path.

Interactions over two years with fellow members of the Solar Radiation Management Governance Initiative (convened by the Royal Society, the academy of sciences for the developing world

Acknowledgements

(TWAS) and the Environmental Defense Fund) have deepened my understanding of the issues considerably. The Royal Society in particular deserves our gratitude for taking the lead globally in investigating and reporting to the public on developments in geoengineering.

I thank them all but implicate none in the final product.

I am grateful to Julian Savulescu for inviting me to visit the Uehiro Centre for Practical Ethics at the University of Oxford, there to present a series of lectures on the subject of this book. My home institution, Charles Sturt University, provides generous support for my work.

I would like to dedicate this book to Joe Romm of Climate Progress for his indefatigable, comprehensive and committed reporting on the science and politics of climate change.

1

Why Geoengineering?

Climate fix

As the effects of global warming begin to frighten us, geo-engineering will come to dominate global politics. Scientists and engineers are now investigating methods to manipulate the Earth's cloud cover, change the oceans' chemical composition and blanket the planet with a layer of sunlight-reflecting particles. Geoengineering – deliberate, large-scale intervention in the climate system designed to counter global warming or offset some of its effects – is commonly divided into two broad classes. Carbon dioxide removal technologies aim to extract excess carbon dioxide from the atmosphere and store it somewhere less dangerous. This approach is a kind of clean-up operation after we have dumped our waste into the sky. Solar radiation management technologies seek to reduce the amount of sunlight reaching the planet, thereby reducing the amount of energy trapped in the atmosphere of 'greenhouse Earth'. This is not a clean-up but an attempt to mask one of the effects of dumping waste into the sky, a warming globe.

Diligent contributors to Wikipedia have listed some 45 proposed geoengineering schemes or variations on schemes. Eight or ten of them are receiving serious attention (and will be considered in the next chapters). Some are grand in conception, some are prosaic; some are purely speculative, some are all too feasible; yet all of them tell us something interesting about how the Earth system works. Taken together they reveal a community of scientists who think about the planet on which we live in a way that is alien to the popular understanding. Let me give a few examples.

It is well known that as the sea-ice in the Arctic melts the Earth loses some of its albedo or reflectivity – white ice is replaced by dark seawater which absorbs more heat. If a large area of the Earth's surface could be whitened then more of the Sun's warmth would be reflected back into space rather than absorbed. A number of schemes have been proposed, including painting roofs white, which is unlikely to make any significant difference globally. What might be helpful would be to cut down all of the forests in Siberia and Canada. While it is generally believed that more forests are a good thing because trees absorb carbon, boreal (northern) forests have a downside. Compared to the snow-covered forest floor beneath, the trees are dark and absorb more solar radiation. If they were felled the exposed ground would reflect a significantly greater proportion of incoming solar radiation and the Earth would therefore be cooler. If such a suggestion appears outrageous it is in part because matters are never so simple in the Earth system. Warming would cause the snow on the denuded lands to melt, and the situation would be worse than before the forests were cleared.

More promisingly perhaps, at least at a local scale, is the attempt to rescue Peruvian glaciers, whose disappearance is depriving the adjacent grasslands and their livestock of their water supply. Painting the newly dark mountains with a white slurry of water, sand and lime keeps them cooler and allows ice to form; at least that is the hope.[1] The World Bank is funding research.

Another idea is to create a particle cloud between the Earth and the Sun from dust mined on the moon and scattered in the optimal place.[2] This is reminiscent of the US military's 'black cloud experiment' of 1973, which simulated the effect on the Earth's climate of reducing incoming solar radiation by a few per cent.[3] Consistent with the long history of military interest in climate control, the study was commissioned by the Defense Advanced Research Projects Agency, the Pentagon's technology research arm, and carried out by the RAND Corporation, the secretive think tank described as 'a key institutional building block of the Cold War American empire'.[4] I summon up the black cloud experiment here to flag the nascent military and strategic interest being stirred by geoengineering. As we will see in chapter 5, the attention of the RAND Corporation has recently returned to climate engineering.

In 1993 the esteemed journal *Climatic Change* published a novel scheme to counter global warming by the Indian physicist P. C. Jain.[5] Professor Jain began by reminding us that the amount of solar radiation reaching the Earth varies in inverse square to the distance of the Earth from the Sun. He therefore proposed that the effects of global warming could be countered by increasing the radius of the Earth's orbit around the Sun. An orbital expansion of 1–2 per cent would do it, although one of the side effects would be to add 5.5 days to each year. He then calculated how much energy would be needed to bring about such a shift in the Earth's celestial orbit. The answer is around 10^{31} joules. How much is that? At the current annual rate of consumption, it is more than the amount of energy humans would consume over 10^{20} years, or 100 billion billion years (the age of the universe is around 14 billion years). This seems like a lot, yet Professor Jain reminds us that 'in many areas of science, seemingly impossible things at one time have become possible later'. Perhaps, he speculates, nuclear fusion will enable us to harness enough energy to expand the Earth's orbit. He nevertheless counsels caution: 'The whole galactic system is naturally and delicately balanced, and any tinkering with it can bring havoc by bringing alterations in orbits of other planets also.'[6]

The caution is well taken, although the intricate network of orbital dependence has stimulated another geoengineering suggestion. The thought is to send nuclear-armed rockets to the asteroid belt beyond the planets of our solar system so as to 'nudge' one or more into orbits that would pass closer to the Earth. Properly calibrated, the sling-shot effect from the asteroid's gravity would shift the Earth orbit out a bit.[7] Of course, if the calibration were a little out, the planet could be sent careening off into a cold, dark universe, or suffer a drastic planet-scale freezing from the dust thrown up by an asteroid strike.

Some of these schemes seem properly to belong in an H. G. Wells novel or a geeks' discussion group, and too much emphasis on them for the delights of ridicule would give a very unbalanced impression of the research programme into climate engineering now underway. That imbalance will be rectified in the next chapters where we will see that serious work is being conducted on schemes to regulate the Earth system by changing the chemical composition of the world's oceans, modifying the layer of clouds that covers a large portion of the oceans and installing a 'solar shield', a layer of sulphate particles in the upper atmosphere to reduce the amount of sunlight reaching the planet. There are some who believe that we will have no choice but to resort to these radical interventions. How did we get to this point? The simple answer is that the scientists who understand climate change most deeply have become afraid.

Hope against fear

In 1959 David E. Price, MD, US Assistant Surgeon General, addressed a conference of industrial hygienists with these words:

> we live under the shadow of a haunting fear that something may corrupt the environment to the point where man joins the dinosaurs as an obsolete form of life. And what makes these thoughts all the more disturbing is the knowledge that our fate could perhaps be sealed 20 or more years before the development of symptoms.[8]

The shadow under which Americans lived was the dual fear of atomic radiation and chemical pollution. Trepidation that the air might be unsafe to breathe gripped the nation. It was the not-knowing that gave rise to a 'mass investment in worry' unmatched, said Price, by an investment in efforts to find out. All that was to change within a few years, spurred by Rachel Carson's earth-shaking book *Silent Spring*, published in 1962, which both confirmed American anxieties about the impact of the chemical war in agriculture and triggered the rise of modern environmentalism.

The haunting fear that something is corrupting the environment has returned, at least for some. Within our breasts fear and hope are duelling. For a few, the reasons to be afraid have prevailed; for most, hope fights on valiantly. Yet hope wages a losing battle; as the scientists each month publish more reasons to worry, and the lethargy of political leaders drains the wellsprings of hope. In 1959 Dr Price invoked that all-conquering sentiment of American greatness, unbounded optimism: 'Stronger than fear is the conviction that what may at times appear to be the shadow of extinction is in reality the darkness preceding the dawn of the greatest era of progress man has ever known.'[9] He was right about the post-war decades. But the world has changed, and now there is a constant trickle of defectors, traitors to hope. To pick out one, the chair of the International Risk Governance Council, Donald Johnston, for ten years the secretary-general of the Organisation for Economic Co-operation and Development (OECD), recently wrote: 'By nature I am not a pessimist, but it requires more optimism than I can generate to believe' that the world will limit warming to 2°C higher than the pre-industrial level.[10] Business as usual is a more likely scenario, he added, taking the concentration of carbon dioxide in the atmosphere from its pre-industrial level of 280 parts per million past its current 395 ppm to 700 ppm this century, 'with horrendous climate change and unthinkable economic and societal consequences'.

The anxiety deepened each year through the 2000s as it became clearer that the range of emissions paths mapped out by experts in the 1990s were unduly optimistic and that the actual growth in emissions, boosted by explosive growth in China, has described a pathway that is worse than the worst-case scenario. When scientists announced that the growth of global greenhouse gas emissions in 2010 was almost 6 per cent, breaking all previous records and wiping out the benefits of a temporary lull due to the global recession, many climate scientists around the world drew a sharp in-breath.

The International Energy Agency of the OECD is a staid organization that for years has shared the worldview of oil and coal industry executives. It is the last international body that could be accused of green sympathies, other than the Organization of Petroleum Exporting Countries. So a frisson of dread ran through the climate change community in November 2011 when the IEA released its annual *World Energy Outlook*, the 'bible' of the energy sector. It exposed the target of keeping warming below the 'dangerous' level of 2°C as a pipe-dream; on current projections, the energy infrastructure expected to be in place as early as 2017 will be enough to lock in future carbon emissions that will warm the Earth by much more. Coal-fired power plants have a lifetime of 50 or 60 years. Waiting for new energy technologies is not an option. If governments do no more than implement the policies they are currently committed to, the IEA expects the world to warm by 3.5°C by the end of the century. 'On planned policies, rising fossil energy use will lead to irreversible and potentially catastrophic climate change.'[11] If those policy goals prove to be more aspirational than actual then the world is on track for average warming of 6°C above pre-industrial levels, which is almost unthinkable.

It's hard to communicate to the public what a world warmed by 3.5°C will be like, let alone 6°C, or even that the IEA, and all the other organizations saying the same thing, should be taken seriously.[12] After all, for many people one unseasonable snowstorm is enough to nullify decades of painstaking scientific study. And psychologists have discovered that, after accounting for all other factors, when people are put in a room and asked about climate change they are significantly more likely to agree that global warming is 'a proven fact' if the thermostat is turned up.[13] Patients with diseases they believe to be serious but untreatable are markedly less likely to agree to diagnostic tests.[14] If it's bad, I don't want to know. Suffice it to say here that 3.5°C means a different kind of world, one hotter than it has been for 15 million years, and not the kind of world on which modern life forms evolved. It would be, eventually, a world without ice – no glaciers, no Arctic sea-ice, no Greenland ice sheet and, almost inconceivably, no Antarctic ice mass. The destabilization of the Earth's climate and natural systems expected this century under the IEA's more 'optimistic' scenario would cascade through the centuries beyond.

For at least a decade, climate scientists and environmental groups have been disturbed by the widening gap between the actions demanded by the evidence and those being implemented or even considered by the major emitting nations. A creeping fear took hold that the truth would be faced too late. After the 1997 Kyoto agreement to reduce global emissions there was an expectation that, having recognized the danger, the world would respond with policies to turn the curve of global emissions downwards. Despite the almost immediate repudiation of the protocol by the United States and Australia it was possible to retain the hope that good sense would prevail. Yet the attacks on the protocol were so persistent and effective that even today journalists unthinkingly reproduce talking points of climate change deniers such as that 'China refused to sign' the treaty. (In fact, China ratified the protocol in August 2002.)

By 2005 the Kyoto Protocol had been ratified by enough nations for it to enter into force. Yet by then it seemed like a pyrrhic victory, its inadequacy highlighted by the fact that growth in world emissions, far from turning down or even stabilizing, had actually accelerated. In the 1970s and 1980s global emissions of carbon dioxide from burning fossil fuels grew at 2 per cent each year. In the 1990s they had fallen to 1 per cent, giving some grounds for cheer. However, from the year 2000, driven mostly by China's astonishing economic expansion, the growth rate of the world's carbon dioxide emissions almost trebled to 3 per cent each year.[15] For those who grasped the enormity of what was at stake, the remnant forces of hope for international action were gathered together for one last mighty push at the Copenhagen conference in 2009. The collapse of the talks left an abyss of despair for the future of the world, one that was not papered over by the milquetoast agreement in Durban in 2011 to begin negotiations for a treaty, to be agreed in 2015, to take effect not before 2020. It is as if the ostriches had awarded themselves another decade to bury their heads. As philosopher René Girard asked: What do we make of today's political leaders 'who claim to be saving us when in fact they are plunging us deeper into devastation each day?'[16]

While governments have been dragging their feet on abatement measures, there has been no shortage of enthusiasm to open up new sources of fossil energy. The Canadian government has facilitated the development of that country's vast tar sands, the most environmentally destructive source of oil. The Russian government, after sending a submarine to plant a flag on the floor of the ice-depleted Arctic sea, encourages its firms to drill for oil, while other oil companies circle. To fend off peak oil (the point after which petroleum production goes into decline because oil fields are being depleted and no new ones can be found), governments in China, South Africa, India and Australia are backing companies that want to revive processes that convert coal into oil. Each of these is worse for the environment than existing sources of fossil fuels, yet they present lucrative commercial opportunities and attract official backing. After pointing out that the amount of carbon in the world's proven coal, oil and gas reserves is five times greater than the amount scientists say it is safe to put into the atmosphere, Bill McKibben notes the irony of US Secretary of State Hillary Clinton travelling to the Arctic to see the damage caused by warming – 'sobering', she called it – before getting down to negotiations with other foreign ministers about how to get access to the new Arctic oil reserves.[17] In this schizoid world, perhaps no nation can compete with Australia. While a modest price was introduced on carbon emissions in 2012, the expansion of new mines to augment the nation's coal exports, already the largest in the world, proceeds apace. According to one estimate, over the next decade the impact on global greenhouse gas emissions of the expansion of Australian coal exports will be 11 times greater than the reduction due to the carbon price legislation.[18]

At the same time, science has come under attack from a well-organized and increasingly vociferous campaign of denial. We will see the contours of this campaign later in the book, but it has taken the form of a flat-out rejection of climate science. News outlets, especially conservative ones, have given prominence to a handful of apparently qualified people who claim to be able to disprove all of the main propositions of climate science. These 'sceptics' have not been able to come up with any evidence for their claims and so they cannot be found in the scientific journals; but that has not dented their appeal to large numbers of lay people, newspaper columnists and political leaders who are looking for a reason, any reason, to reject the vast accumulation of evidence from a range of sources showing that we are in deep trouble.

In a question and answer session following a public lecture, the prominent (and genuinely sceptical) climate scientist Chris Rapley was vociferously challenged by a climate denier in the audience. (The individual's wife fled the lecture theatre as he rose to speak!) After responding calmly to a torrent of accusations, to no effect, Rapley stopped and asked his accuser what it would take to convince him that he was wrong, that climate change is real, dangerous and caused by humans. His critic ignored the question and it was clear to the audience that no amount of evidence could change his mind. A fair-minded man, Rapley later posed the same question to himself. He answered that he would change his mind in response to a research paper, published in a peer-reviewed journal, revealing a feedback effect that neutralized climate change, along with an explanation as to why it had remained undetected or latent until now. The new evidence would require confirmation from an expert in the field whom he holds in esteem.[19]

A sceptic is one who carefully filters received knowledge to see which propositions stand up to independent scrutiny. But one thing we immediately notice about the contributions of climate 'sceptics' is the absence of a quizzical, thoughtful approach. Among those who debate the science of climate change they are the ones who profess to be most *certain*, insisting vehemently on the falsity of the claims of climate scientists and convinced of the correctness of their own opinions. The true sceptics are, of course, to be found among climate scientists themselves. As a matter of cultural practice and professional rivalry, research scientists routinely subject the work of their peers to the most critical scrutiny. It is a mark of quiet professional pride to find mistakes in the work of one's fellow researchers. If the Intergovernmental Panel on Climate Change (IPCC) of the United Nations can be accused of anything, it is of an excess of caution in reporting the science.

Feedback science

While climate scientists observed these baleful political developments, their work provided additional grounds for disquiet. Building on the discoveries of palaeoclimatologists and more advanced knowledge of the functioning of the Earth system, they began to focus on the dangers of feedback effects in the climate system, that is, responses in the Earth system that amplify or dampen the direct effects on warming of rising greenhouse gas emissions. For example, as warming melts the Arctic ice cap (which coats the Arctic Sea) the exposed water is darker than the highly reflective ice it replaces and absorbs more heat from the

Sun. 'Arctic amplification' has seen the rate of warming in the Arctic occur at two to four times the global average.[20] Many in the expert community were shocked by the dramatic declines in Arctic summer sea-ice in 2005 and especially 2007. Warmer Arctic waters are causing complex changes to climate patterns in the northern zones, including melting of permafrost (now a misnomer). The release of frozen methane, a highly potent greenhouse gas, is expected to further amplify warming.

There are negative feedback effects that dampen warming and tend to return the climate system towards an equilibrium state – for example, over very long timescales enhanced chemical weathering of rocks may see more carbon dioxide taken out of the atmosphere and stored in the deep ocean – but overall the destabilizing effects are expected to be much more powerful.[21] Since the early 2000s research into feedback effects has gathered pace, not least because understanding these processes is essential to filling the gaps between the climate models and the actual behaviour of the climate system.

The study of feedbacks has been closely related to another emerging idea – that of tipping points. For example, when warming in Siberia reaches a certain threshold the frozen ground will thaw, releasing methane into the atmosphere. The Earth's climate is a 'non-linear' system, that is, changes in one variable do not lead to simple proportional changes in related ones. The equations are far more complex. In non-linear systems, a small change in one state may initially have only small effects but at some point a threshold may be crossed where the system, driven by amplifying feedbacks, flips suddenly into a new state. Research emerging from palaeoclimatologists has fed these concerns. They have discovered many instances in the Earth's climate record of the climate shifting abruptly from one state to another within a few decades. The esteemed palaeoclimatologist Wally Broecker highlighted this fact when in 1995 he wrote: 'The palaeoclimate record shouts out to us that, far from being self-stabilizing, the Earth's climate system is an ornery beast which overreacts even to small nudges.'[22]

The existence of tipping points destroys the comforting idea that the slow build-up of greenhouse gases is causing a gradual change in temperature and that when it gets bad enough we can do something about it. The essential belief on which global negotiations were founded was increasingly seen to be dangerously wrong. The emerging science of abrupt climate change was reviewed in a landmark report, published in 2002 by the US National Research Council.[23] One of the authors, the director of the Woods Hole Oceanographic Institution, noted 'recent and rapidly advancing evidence that Earth's climate repeatedly has shifted abruptly and dramatically in the past, and is capable of doing so in the future.' Dr Robert Gagosian went on:

This new paradigm of abrupt climate change has been well established over the last decade by research of ocean, earth and atmosphere scientists at many institutions worldwide. But the concept remains little known and scarcely appreciated in the wider community of scientists, economists, policy makers, and world political and business leaders. Thus, world leaders may be planning for climate scenarios of global warming that are opposite to what might actually occur.[24]

The idea was born that within the next few decades we may face a 'climate emergency'. Palaeoclimatologists explained that although the Earth's climate has always been in a state of flux, shifts may be so sudden that natural systems, such as forest ecosystems, are unable to adapt and thus disappear. Abrupt climate change in the past is thought to explain some mass extinctions. In 2009 a group of eminent Earth scientists summarized their growing concerns about feedback effects, tipping points and abrupt climate change in an article in *Nature*. Current climate models, they wrote:

> do not include long-term reinforcing feedback processes that further warm the climate, such as decreases in the surface area of ice cover or changes in the distribution of vegetation. If these slow feedbacks are included, doubling CO_2 levels gives an eventual temperature increase of 6°C (with a probable uncertainty range of 4–8°C). This would threaten the ecological life-support systems that have developed in the late Quaternary environment [the last half to one million years], and would severely challenge the viability of contemporary human societies.[25]

The floodgates

In the face of ever-increasing global greenhouse gas emissions, political inertia and worries about sudden climate change, some scientists began to mull over what could be done to slow the world's apparently unstoppable rush into the abyss. Among themselves they began to talk about possible responses to a climate emergency, such as a massive methane release following accelerated melting of permafrost, the collapse of the West Antarctic ice sheet, or rapid disappearance of the Amazon forests due to heat-stress and drought. Any of these could quickly shift the global climate into a new state, and there would be no way of recovering the situation. How could we intervene to prevent these things happening? If Plan A, persuading the world to cut emissions, is failing, shouldn't we have a Plan B? The search for an alternative to emission cuts led to the idea of engineering the climate.

In the 1990s proposals for geoengineering were regarded by the mainstream as fanciful and a distraction from the real task of reducing emissions. Although Plan B had been a topic of private speculation for some years, almost all climate scientists took the view that the availability of an alternative to cutting emissions, even if manifestly inferior, would prove so alluring to political leaders that it would further undermine the will to do what must be done. To canvass climate engineering, let alone advocate it, would be unethical. But the longer political leaders prevaricated the louder the silence surrounding geoengineering became. The frustration became too much for Paul Crutzen, the eminent Dutch atmospheric scientist who had shared the Nobel Prize for discovering the key chemical reactions needed to explain the hole in the ozone layer. So he penned an editorial essay, 'Albedo enhancement by stratospheric sulfur injections: A contribution to resolve a policy dilemma?', published in the journal *Climatic Change* in 2006.[26] His intervention broke the taboo on geoengineering.

Expecting the political process to respond adequately to the imperative to cut emissions, Crutzen argued, had become a 'pious wish'. It would be prudent to invest in a substantial research programme to test the feasibility of cooling the Earth by injecting sulphate aerosols into the upper atmosphere in order to reflect a greater portion of sunlight back into space. Crutzen expressed particular concern at the 'Catch-22' presented by the fact that governments in developing countries are following industrialized countries with measures to clean up urban air pollution from cars, factories and power plants, responsible, he wrote, for some 500,000 premature deaths each year. That pollution, especially the high sulphur emissions over much of East Asia, is helping to cool the planet; cleaning up the air would, over a brief decade, lead to an unprecedented increase in global temperature by almost 1°C over land, and 4°C in the Arctic. Without an 'escape route against strongly increasing temperatures', he wrote, continued emissions growth combined with anti-pollution laws would bring about potentially catastrophic effects on ecosystems. Noting that the development of the Antarctic ozone hole was 'sudden and unpredicted', Crutzen wanted to alert the world to the risks of unexpected warming.

Many of Crutzen's colleagues at the Max Planck Institute and elsewhere reacted angrily to his intervention. In anticipation, one of his associates, Mark Lawrence, wrote a paper in his defence titled 'The geoengineering dilemma: To speak or not to speak?'. Lawrence referred to the 'passionate outcry by several prominent scientists claiming that it is irresponsible to publish' calls for research into geoengineering, and provided several counter-arguments for why it was time to break the taboo.[27] Nevertheless, the ferocity of the response shocked Crutzen. He weathered the storm and time presently proved that if he had not intervened someone else would have soon enough; the pressure had become irresistible.

By early 2009, three years after Paul Crutzen opened the floodgates, more than half of leading scientists who responded to a poll by the *Independent* newspaper agreed that 'the situation is now so dire that we need a backup plan'.[28] That was before the Copenhagen fiasco. A third disagreed with the proposition, not because they assessed the situation differently but because they believed the better response is to commit more strongly to Plan A. The Copenhagen conference in December 2009 was the first of the annual international climate change jamborees at which geoengineering proposals had a significant presence at various side events.[29] A year later the IPCC decided for the first time to incorporate into its next report an evaluation of geoengineering as a response to global warming.

Research into various schemes to engineer the climate has been accelerating rapidly. A network of scientists, entrepreneurs and advocates has formed and is gaining influence in the scientific community and in government. According to one observer, John Vidal:

> From just a few individuals working in the field 20 years ago, today there are hundreds of groups and institutions proposing experiments. ... The range of techno-fixes is growing by the month ... Most are unlikely to be considered seriously but some are being pushed hard by entrepreneurs and businessmen attracted by the potential to make billions of dollars in an emerging system of UN global carbon credits.[30]

When this was written in 2011, I think the first claim was something of an overstatement, although it will be true soon enough. While the number of researchers expressing interest in the area has grown substantially, and entrepreneurs and scientists are registering patents for various techniques, the international debate over geoengineering and its governance remains dominated by a very small group of experts, mostly scientists but including a handful of economists, lawyers and policy experts. In 2009 some members of that small group could write: 'Nearly the entire community of geoengineering scientists could fit comfortably in a single university seminar room, and the entire scientific literature on the subject could be read during the course of a transatlantic flight.'[31] That was an exaggeration then and is certainly untrue now as the scientific literature has ballooned.

That someone of Paul Crutzen's stature and undoubted commitment to protecting the natural world – he was described in *Time* magazine as 'the chief scientific caretaker of life on the planet'[32] – should call for serious research into geoengineering as a response to global warming must give pause for thought. Geoengineering presents a profound dilemma, not just for climate scientists, but also for environmentalists. It is a dilemma that all citizens will soon need to face. Many find repellent the idea, embodied in some geoengineering schemes, of attempting to take control of the Earth's climate as a whole. It is, surely, the ultimate expression of humankind's technological arrogance. Yet if the alternative is to stand back and watch humanity plunge the Earth into an era of irreversible and hostile climate change, what is one to do?

Perhaps Crutzen's only offence was to arrive at the conclusion a decade ahead of most others. On the other hand, his well-meaning intervention might legitimize the stance of hitherto fringe voices whose motives are less politically pure or sympathetic to environmental protection. That was his colleagues' fear, and it was a reasonable one. As we will see, climate engineering is intuitively appealing to a powerful strand of Western technological thinking and conservative politicking that sees no ethical or other obstacle to total domination of the planet. It is a Promethean urge named after the Greek titan who gave to humans the tools of technological mastery. Promethean plans have always met resistance from those who share a deep mistrust of human technological overreach, those who heed the warning that Nemesis waits in the shadows to punish Hubris. If Prometheus is the god of technological mastery, who is the Greek divinity of caution? Perhaps the closest is Soteria, the goddess of safety, preservation and deliverance from harm.[33] I will suggest that climate engineering is the last battle in a titanic struggle between Prometheans and Soterians, with the prize nothing less than the survival of the world we know now.

As will become apparent, one cannot assume a simple correspondence between Promethean and Soterian sympathies and support for and opposition to geoengineering. Paul Crutzen, for example, is a Soterian. As will become apparent, I have serious doubts about the wisdom of any attempt by humans to take control of the weather. The reasons will become plain, but at their heart is a conviction that the Earth is unlikely to collaborate in our plans, and we should heed the kind of warning most famously expressed by Robert Burns:

> The best laid schemes of Mice and Men
> oft go awry,
> And leave us nothing but grief and pain,
> For promised joy!

I hope to explain, not least by drawing on Earth system science, an understanding of the Earth that inclines to this conviction.

Yet if I am not *for* geoengineering then that means I must accept climate disruption, doesn't it? If most of the world continues to entertain the fantasy that global warming is trivial or a long way off, or that governments will respond in time to avoid climate chaos, and if Crutzen and a few others, despairing at this blindness, want to be ready to intervene radically when the world comes to its senses and realizes cutting emissions will come too late, where does that leave me politically and philosophically? Answering that question is a work in progress, one I hope will be resolved by the time I reach the last chapter of this book.

2

Sucking Carbon

The great carbon cycle

Geoengineering methods are typically divided into two types. Carbon dioxide removal methods aim to extract the gas from the atmosphere and deposit it somewhere safer; as we will see, they variously identify these storage options as the soil, vegetation, the oceans and back underground. They would work by manipulating one of the great natural processes that makes the Earth a dynamic evolving entity, the global carbon cycle, which continually exchanges carbon between the atmosphere, the oceans and the biosphere (and, much more slowly, the lithosphere). The second type, solar radiation management (considered in the next chapter), aims to cool the planet by reflecting a greater proportion of incoming radiation from the Sun back out to space. While carbon dioxide removal methods target the source of the malady – too much carbon in the atmosphere – solar radiation management methods target one of its symptoms: too much heat.

The usual distinction between geoengineering methods conceals as much as it reveals. Although all aim to alter the global climate, perhaps a more useful division would be between those that aim to intervene in the functioning of the Earth system as a whole, where the risks are greater, and more localized interventions that have only regional environmental impacts, where the costs of failure are lower. Nevertheless, the usual distinction also points to an important difference – whether the intervention targets the disease or only a symptom of it – and for that reason I stay with it in this and the next chapter to describe how they work. However, when it comes to the larger questions of geopolitics and ethics, and what climate engineering can tell us about humans in the twenty-first century, it will be more enlightening to focus on the system-altering technologies rather than localized ones.

Fossilized carbon is congealed solar energy, deposited millions of years ago when massive numbers of dead organisms were transformed by heat and pressure beneath layers of rock. When we extract fossil fuels from coal mines, oil wells and natural gas deposits, and burn them for their captured energy, the carbon atoms combine with oxygen and float into the atmosphere as carbon dioxide. By absorbing more heat near the Earth's surface, the atmosphere enriched by carbon dioxide causes global warming. But then what happens to the carbon? As it circulates around the globe, some carbon dioxide is absorbed by land-based plants and microorganisms in the soil. Some is absorbed by the oceans. In fact, over the last decade or so the biosphere and the oceans have each absorbed a quarter of our emissions.[1] But that is only the beginning of the story.

In the case of the terrestrial biosphere, vegetation and other life forms are in a constant flux of growth and decay. At times the flows do not balance each other out. In recent decades, the net amount of carbon stored in the Earth's soils and vegetation has been gradually rising; despite continued deforestation in parts of the world, the take-up has been greater than the release. But this can only be temporary as there is a limit to the capacity of the biosphere to absorb carbon, a limit that will decline as more land is turned over to farming, as trace nutrients are depleted and as climate change advances. In future decades we cannot rely on the terrestrial biosphere to soak up much, if any, of our extra emissions; indeed, it may well become a net source of emissions. Growing trees is good, but it cannot save us from climate change.

So the capacity of the world's oceans to absorb carbon dioxide is of decisive importance to the future climate. How does it work? Carbon dioxide from the air initially dissolves into the top layer of the ocean, more so in choppy turbulent seas. But the top layer is saturated and can absorb only as much carbon dioxide as is drawn down into the deep layers, layers that are not well mixed and so can take up more carbon dioxide. Cold water can absorb more carbon dioxide than warm water so the cold ocean layers of the high latitudes (towards the poles) do most of the work, even though they account for only 2–3 per cent of the Earth's surface.[2]

However, as more carbon dioxide is absorbed the surface layers of the oceans become more acidic (mixing carbon dioxide with water produces carbonic acid), slowing their ability to take up our carbon dioxide emissions. And as the globe warms so do ocean waters, further reducing their capacity to soak up more carbon. Nevertheless, over decades and centuries the atmosphere and oceans will continue to exchange carbon dioxide in a process of 'equilibration'. So over the long term some 70–75 per cent of this century's carbon dioxide emissions will eventually be absorbed by the oceans, with some 20–25 per cent remaining in the atmosphere. Some of the increased carbon dioxide 'stored' in the atmosphere will stay there for many centuries, long after the last tonne of fossil carbon has been shovelled into the furnace of a coal-fired power plant. Over an even longer time-scale, the excess carbon dioxide very gradually penetrates the ocean depths, slowly drawing down the atmospheric content over thousands of years. Even so, 10 or 12 per cent of our fossil fuel emissions would persist in the atmosphere after 10,000 years.[3] As far as scientists can estimate, if we released all of the fossil carbon, the Earth would remain 3–5°C warmer in 10,000 years' time, having peaked at around 8°C hotter in a century or two.[4]

From table 1 we can see where the Earth's carbon is today, and how humans have shifted carbon around the planet over a century and a half of industrial and agricultural activity. What flows from one reservoir must go to another, so the right-hand column adds up to zero. It is apparent that we have redistributed a substantial amount of carbon from its ancient storehouse under the ground into the atmosphere and the oceans. What leaps out from the table is the fact that by far the largest share of carbon on planet Earth is

Table 1 Where is the Earth's carbon stored?

Carbon reservoir	Pre-industrial amount stored (GtC)	Change from pre-industrial times (1850 to end 2010) (GtC)
Atmosphere	590	+219
Land (vegetation, soils)	3,800	−16
Fossil carbon	>6,000	−363
Ocean (surface, intermediate and deep)	38,000	+160

Note: GtC = gigatonnes (billions of tonnes) of carbon. The land change is the sum of uptakes to land (+137 GtC) and cumulative emissions from land use change (−153 GtC).

Sources: Pre-industrial stocks from C. Sabine et al., 'Current status and past trends of the global carbon cycle', in C. Field and M. Raupach (eds), *The Global Carbon Cycle: Integrating Humans, Climate, and the Natural World* (Washington, DC: Island Press, 2004), pp. 17–44. Changes in stocks from C. Le Quere et al., 'Trends in the sources and sinks of carbon dioxide', *Nature Geoscience*, 2 (2009), as updated in Global Carbon Project, at http://www.globalcarbonproject.org/carbonbudget/index.htm (accessed Feb. 2012). Mike Raupach provided data in a more suitable format.

stored in the deep and intermediate layers of the oceans (the surface layer holds relatively little) – ten times more than is stored underground as coal, oil, natural gas and peat. The table does not show the enormous amount of carbon stored in carbonate rocks in the lithosphere because that carbon is so firmly fixed that over human timescales it hardly changes at all.[5] The carbon stored in rocks is effectively immobile while carbon in the atmosphere, the biosphere and, less so, the oceans is highly mobile.[6]

It is immediately apparent why getting more carbon more quickly into the oceans is so alluring. As the quantities there are so vast our additions would not seem to make much difference. However, as we will see, things are not so simple.

It is not feasible to describe here all of the proposed methods of climate engineering, so in this chapter I concentrate on those carbon dioxide removal methods that are attracting most attention from researchers and investors or seem more likely to attract the interest of policy-makers. They are ocean fertilization, liming the oceans, enhanced weathering of rocks, air capture and some ideas for bio-geoengineering, including carbon sequestration in soils. I give more space to those technologies that are receiving more attention from researchers and which I judge to be most likely to prove attractive to implement at some stage in the future. Each technology has its own champions and some will disagree with my emphasis.

A few overviews of the technologies are available but to date there is no book that provides a thorough explanation of them for a lay audience. This chapter and the next are necessarily somewhat technical. My aim is to explain it all in a way that can be understood by the reader whose scientific education stopped, like mine, before university. Even so, the less interested reader may prefer to skim them. The essential message is that when we mess with ecological systems things soon become much more complicated than they first seem, and as the complications multiply so do the uncertainties and the dangers.

Fertilizing the ocean

Compared to around 800 billion tonnes of carbon occurring in the atmosphere, and almost 4,000 billion as organic matter in the biosphere, the quantity of carbon sequestered in the deep ocean is enormous.[7] The 38,000 billion tonnes stored in the oceans makes the annual emissions by humans into the atmosphere, 10 billion tonnes, appear paltry. Why, scientists are asking, can we not find a way to send some or all of our annual emissions into the ocean depths, where they would be barely noticed?

One way to do so might be to capture our carbon dioxide waste somehow and inject it into the deep ocean using pipes. That has been proposed, but more interest has been devoted to finding ways to accelerate the natural mechanism by which carbon finds its way to the deep ocean. The effect would be to hasten transfer of carbon from coal mines and oil fields via the atmosphere into the watery depths. How might we enhance the natural process of sending carbon to the deeps? The answer might lie in what is known as the marine biological pump.

Nutrients continuously circulate through the oceans, not only horizontally but vertically through a constant mixing of water layers. In the upper layer (the top 200 metres) turbulence due to winds and currents mixes carbon dioxide from the air into surface waters, as bubbles which then dissolve. Tiny marine plants known as phytoplankton absorb dissolved carbon dioxide, minerals and sunlight, and multiply into 'blooms' of growth. On death, gravity causes the plankton to sink. But not much of their carbon reaches the bottom, as on the way down most is oxidized into carbon dioxide again by decomposer organisms. Sooner or later it is returned to the surface waters by upwelling currents. However, phytoplankton are only part of the marine food web. Some are devoured by larger sea creatures, whose excrement and carcasses add to the 'marine snow' of organic detritus falling to the deep ocean. This is shown in figure 1.

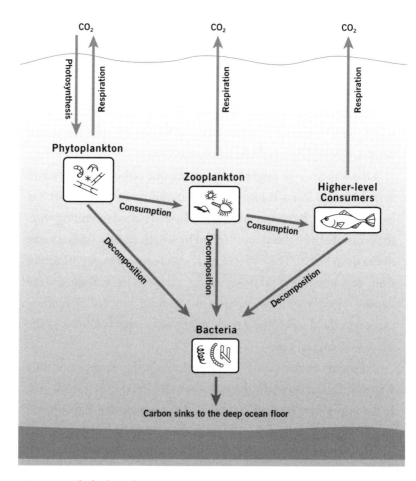

Figure 1 The biological pump

Unless the depths are to become increasingly saturated with carbon from the marine snow, there must be a process that brings some of it back to the surface. Over very long periods, carbon-rich deep currents of cold water eventually find their way to the surface, where some of their carbon is oxidized and released into the air as carbon dioxide. This may take centuries or millennia. Deep currents can move so slowly that some waters have not felt the sunlight of the 'photic zone' for a thousand years or more.

The rate at which carbon is 'pumped' down to the intermediate and deep ocean layers by gravity depends on how much organic material is being created in surface waters (known as ocean productivity), which in turn depends on the suitability of conditions for marine life. Among those conditions are the availability of macronutrients – phosphorus, nitrogen and carbon itself ('macro' because organisms need a lot of them) – and micronutrients, especially iron. It occurred to scientists that if a shortage of one of these was limiting plankton growth in an area of ocean, then perhaps the artificial addition of the missing ingredient could stimulate algal blooms. So the idea of ocean fertilization was born.

Observation of natural events has reinforced the idea. Satellites can now send us images of green swaths of phytoplankton forming across oceans after dust storms have dumped iron-rich dust from land. Intriguingly, the effect can be so powerful that iron concentrations in the oceans have been linked to ice ages. Ice-age peaks (known as glacial maxima) are associated with atmospheric carbon dioxide concentrations 80–100 parts per million lower than the pre-industrial average of 280 ppm. (Today the concentration has risen to 395 ppm.) It's estimated that around a third of that decline of 80–100 ppm was due to an enhanced biological pump associated with elevated iron in the oceans.[8] The origins of the additional iron are uncertain, although for some time palaeo-oceanographers have known from ice cores that huge algal blooms have occurred in the Southern Ocean after natural injections of iron sourced from dust blown in from continents to the north (Africa or South America) or from the melting of icebergs, iron-enriched by dust storms, calved from South American glaciers.

Enthusiasm for iron fertilization was stoked by laboratory studies suggesting that one tonne of iron added to the ocean in a suitable form could remove 30,000 to 110,000 tonnes of carbon from the atmosphere. As further research revealed more complex ecosystem effects, however, estimates of the sequestering potential of added iron fell sharply. Zeal has been dulled, but the prospects remain high enough for research to continue. The evidence suggests that the efficiency of the biological pump (its effectiveness in taking carbon to deep layers) varies widely, from 20 per cent near Hawaii to 50 per cent in the cold waters of the North Pacific. One important factor is the type of phytoplankton whose growth is boosted. Diatoms, a type of hard-shelled zooplankton, perform well, as they are less favoured by predators and sink more quickly, before they are oxidized by decomposition back into carbon dioxide. One of the few predators of hard-shelled diatoms are gelatinous animals of 1–10 centimetres known as salp. These 'jelly balls' can vacuum up huge quantities of phytoplankton and excrete 'large, heavy, carbon-rich fecal pellets that sink much faster than the feces or dead bodies of other zooplankton.'[9] So it's suggested that phytoplankton blooms combined with salp colonies may be particularly effective at getting carbon out of the air and down to the bottom of the sea.

Yet already we begin to feel uneasy about this kind of enterprise. If we have learned anything from the science of ecology it is that interventions in natural systems often turn out to be more complicated the more they are scrutinized. And they often reveal counter-intuitive effects. We know that natural processes on land are never simple; in the oceans ecological and bio-geochemical dynamics are equally intricate. For a start, ocean ecosystems have three dimensions. And while the land stays in one place, the oceans are constantly on the move. Plankton (the name derives from the Greek word 'wandering', because they have no propulsion system but drift in surface currents) form the bottom rung of the marine food chain; any major change will cascade through the ecological system, often in unexpected ways. As we saw, a bloom of phyto-plankton (plants) is typically set upon immediately by tiny zooplankton (animals). Krill, fish, whales and other marine crea-tures join the feast. Much of the carbon fixed in the phytoplankton does not find its way to the ocean floor but circulates in the surface waters before being emitted as carbon dioxide back into the atmos-phere. But some does make it to the intermediate ocean layers – where it may remain for decades before surfacing – and to the deep ocean, known as the abyssal zone.

But that is only the beginning of the ecological ramifications. An algal bloom fertilized by an artificial injection of iron soaks up large amounts of other nutrients, like phosphate, nitrate and silica, depleting surrounding waters. While iron fertilization stimulates biological productivity in one area, macronutrient stealing can see it fall in others. Making one pump work harder slows down others. As one expert said: 'you might make some of the ocean greener by iron enrichment, but you're going to make a lot of the ocean bluer'.[10] And those less productive blue patches will send less carbon to the depths, so it is a self-compensating system.

By distorting the bottom of the food chain, iron fertilization naturally affects its composition further up. Stocks of some fish species will swell (which may or may not be a good thing for humans) and some whale populations too. When they eat phytoplankton, krill concentrate iron in their bodies. When whales, which devour huge quantities of krill, swim on and excrete their iron-rich faeces they fertilize the oceans elsewhere and stimulate further algal growth.[11] So iron fertilization may see whale numbers expand. Hunting has caused a crash in the populations of some whales in the Southern Ocean, which has in turn depleted iron levels and thus that ocean's ability to soak up carbon dioxide from the atmosphere. So one of the benefits of ending whaling may be to enhance the biological pump.

In short, fertilizing the ocean on a scale large enough to extract substantial amounts of carbon dioxide from the atmosphere would bring about large but little understood changes in planktonic communities and the wider ocean ecosystems that sustain them.[12]

There are further effects. By turning the ocean green, plankton blooms may block sunlight needed for growth by corals, although iron fertilization is unlikely to be tried in waters near coral reefs. Greener ocean surfaces are also less reflective and so may absorb more solar radiation, warming that part of the seas and changing circulation patterns. Against this, and to complicate matters considerably, plankton blooms produce a gas called dimethyl sulphide, a sulphur compound that rises into the atmosphere. There it oxidizes to form sulphate aerosols, tiny particles that in turn supply the nuclei around which water vapour condenses to form clouds. The oceans naturally give off huge amounts of dimethyl sulphide, which has been called the 'smell of the sea'.[13] (Astonishingly, it is thought some seabirds use its odour to find rich feeding areas.) Dimethyl sulphide from phytoplankton is believed to play an essential role in the regulation of the Earth's climate because cloud cover affects the Earth's albedo or reflectivity. So by stimulating the oceans to give off more of this gas, iron fertilization may increase cloud cover and have a cooling effect. Indeed, some geoengineering schemes, considered in the next chapter, are designed to do precisely that – increase low-altitude marine cloud. However, our understanding of the dense network of links between ocean biology, atmospheric chemistry, cloud physics and Earth's albedo is in its infancy.

Since the early 1990s scientists have carried out a dozen or so iron fertilization experiments. The waters best suited to iron fertilization occur in the Southern Ocean. The Southern Ocean is dominated by the massive Antarctic Circumpolar Current (figure 2). Linked to all other major ocean currents, the Antarctic Circumpolar Current follows a 20,000 kilometre path around the Antarctic ice mass, moving west to east at a dawdling one knot (a

Figure 2 Southern Ocean circulation

little less than 2 kilometres per hour). Four kilometres deep and 100–200 kilometres in breadth, it carries water equivalent to 150 times the volume of all of the world's rivers combined.[14] The Southern Ocean is of enormous importance to the global climate and the carbon cycle, although it is not well understood.[15] While the least polluted in the world, over recent decades its waters have become warmer, less salty and more acidic. The Southern Ocean is a gigantic carbon store, responsible for soaking up around 40 per cent of anthropogenic carbon dioxide absorbed by all the oceans.[16] It has also absorbed a disproportionate amount of the Earth's additional heat due to global warming.

The Southern Ocean is attractive as a site for accelerating the biological pump because the potential productivity due to high levels of macronutrients is constrained by low levels of iron. It is easy to become excited from the deck of a research vessel as iron slurry pumped in its wake almost immediately changes the colour and even the smell of the seas. But in the cool light of analysis the results have been disappointing. While it is straightforward to increase biological productivity in the surface waters, getting the carbon down into deep waters for more long-term storage meets obstacles. Measured at a depth of 200 metres, each tonne of iron seems to sequester only around 200 tonnes of carbon, a tiny fraction of the new productivity that blooms on the surface, and a far cry from the theoretical 30,000–110,000 tonnes. Initial estimates suggesting that a programme of iron fertilization could remove as much as a billion tonnes of carbon from the atmosphere each year – around one-tenth of current human emissions – now appear optimistic. It would require assigning all of the Southern Ocean to a continuous process of iron fertilization. The flow of carbon would then be from coal mines and oil wells into the atmosphere (via power plants and vehicles) and then into the artificially created 'plughole' in the Southern Ocean, ideally ending up in the ocean depths.

Experiments in iron fertilization alter the chemical composition of perhaps a hundred square kilometres of ocean over several days. The results provide few clues as to the implications of altering the chemistry of an area big enough to change atmospheric carbon dioxide concentrations measurably. Nevertheless, using the highest estimates for the rate at which carbon dioxide could be soaked up from the air and for the rate at which carbon could sink to the bottom, cumulative sequestration from a massive fertilization effort over 100 years is in the range of 25–75 billion tonnes of carbon,[17] compared with expected cumulative emissions from fossil fuel combustion of 1,500 billion tonnes under business as usual, around 3 per cent. In the meantime, ocean acidification and temperatures would reach a level at which algal populations would be severely reduced. This is one reason why climate engineering without emissions cuts would be disastrous.

Oceans are in continuous flux; even at abyssal depths, storage is not permanent. Still waters run deep; but even at its deepest, seawater is not still. After around 100 years, three-quarters of the carbon exported to the ocean floor is returned to the surface by upwelling currents.[18] However, the iron in the dead organic material is decoupled from the carbon and other nutrients and mostly stays in the depths, so if the carbon that has resurfaced is to be absorbed once more by phytoplankton the ocean will need to be fertilized with iron again.

It may have occurred to the reader that manipulating the global carbon cycle through promoting marine life must also disrupt other processes on which life depends, notably the great cycles that distribute phosphorus and nitrogen through the Earth system. Phosphorus, an essential nutrient, is cycled through the biosphere (living things), the hydrosphere (water) and the lithosphere (the Earth's crust) by bio-geochemical processes. The oceans are integral to the slow cycling of phosphorus through living systems. It might be thought that phosphorus-deprived oceans could be augmented artificially. However, with the use of phosphoric fertilizers in agriculture humans are already washing nine to ten times more phosphorus into the oceans than occurs naturally, promoting algal blooms and expanding the area of the world's oceans starved of oxygen.[19] Moreover, supplies of phosphorus (on which agriculture depends) are limited, with some experts predicting 'peak phosphorus' in 2030. In addition, studies point to human intervention already being responsible for 'a massive disruption of the global nitrogen regime'.[20] Disturbances to the phosphorus and nitrogen cycles have been named among the nine 'planetary boundaries' the transgressing of which 'may be deleterious or even catastrophic' due to the high risk of triggering abrupt environmental change at a planetary scale.[21]

Hopes that iron fertilization could provide a substantial response to global warming waned in 2009 after a three-month experiment carried out in the Southern Ocean by a team of German and Indian scientists.[22] (Perhaps dreading 40 days of German food, the Indians took their own spice supply and cook.) The experiment, known as LohaFex, was embroiled in controversy before it began, and the voyage was stalled while the German environment and research ministries arm-wrestled over its legality. Eventually the ship set sail for the southern Atlantic, searching for a suitable eddy – a rotating water column with a diameter of around 100 kilometres that would provide a well-defined zone for the experiment. After 4 tonnes of iron dust, dissolved on board with seawater, had been spread over a 300 square kilometre patch of ocean, phytoplankton quickly bloomed, with instruments showing carbon dioxide being drawn from the waters. Against expectations, however, the bloom's growth stopped after two weeks. The algae had attracted abundant predators, mainly in the shape of small crustacean zooplankton known as copepods (meaning 'oar-feet'). The ecosystem went into recycling mode, with the carbon dioxide cycling through the planktonic life forms in the upper ocean zone, so that very little of the bio-captured carbon made it to the deep.

The LohaFex scientists realized that the problem could be traced to the fact that waters in that part of the Southern Ocean contain no silicon, the element used by diatoms to make their glassy shells, shells that deter predators so that the carbon can find its way to the ocean floor. They speculated that the silicon had been extracted in the past by natural blooms fertilized by dust storms from Patagonia or by melting icebergs. Instead of diatoms a host of softer plankton flourished and they were immediately eaten by a range of predators before they'd sunk a few metres. A previous study had found that iron fertilization could take more carbon dioxide to the deep ocean if carried out in a region sympathetic to diatom blooms.[23]

It turns out that two-thirds of the Southern Ocean has very little silicon so iron fertilization would be ineffective at sending carbon to the ocean depths. Efforts would need to be concentrated in those areas where fertilization would favour diatoms. Solving humanity's global warming problem seems a heavy burden for these tiny marine creatures. The researchers estimate that if the silicon-rich third of the Southern Ocean were seeded with iron, the biological pump would at most take 1 billion tonnes of carbon dioxide from the atmosphere each year. On that basis, the deployment of full-scale iron fertilization would see a third of the Southern Ocean – around 5 per cent of the Earth's surface – serve as a sink for a tenth of the world's current annual excess carbon dioxide emissions.

Liming the seas

Iron fertilization is but one method aimed at getting carbon out of the atmosphere and into the oceans. Instead of attempting to exploit fickle biological processes in the ocean, another method would exploit well-known and simple chemical reactions. As we saw, the surface layers of the oceans, whipped up by winds, interact with the atmosphere and the mixing allows carbon dioxide to pass from one to the other. In some conditions there may be a net absorption of carbon dioxide into surface waters, in others a net desorption. Cold water can absorb more carbon dioxide, but when upwelling currents bring warm water to the surface carbon dioxide is released back into the atmosphere. Unless disturbed, a natural back and forth process between atmosphere and ocean maintains a balance over time.

When humans increase the concentration of carbon dioxide in the atmosphere by burning fossil fuels the oceans absorb more carbon, which increases their acidity. In fact, the oceans are absorbing a quarter of anthropogenic emissions; without this effect global warming would have been faster. But when combined with seawater, carbon dioxide creates carbonic acid, a weak acid that reduces the oceans' alkalinity. Acidification slows down the biological pump (because many marine creatures cannot cope), and warmer oceans from the enhanced greenhouse effect make carbon dioxide less soluble. Both processes limit the ocean's ability to take up more carbon dioxide. Acidification is also worrying because it inhibits coral growth and shell formation by all sorts of sea creatures. If it continues, it will severely disrupt marine ecosystems. In the absence of major emission cuts, or other offsetting measures, coral reefs are expected to come under severe stress within decades.[24]

One approach that is garnering interest – first put forward by Haroon Kheshgi[25] – aims to counter the acidification of the oceans by sprinkling lime (calcium oxide) over the oceans. The idea is simple: enhance the ocean's ability to absorb carbon dioxide by dispersing lime, an alkali, so as to return its alkalinity to normal levels. While iron fertilization relies on marine biota and the biological pump, which we have seen is a complex and uncooperative beast, enhancing ocean alkalinity by adding lime relies on some simple and well-established chemical reactions.

If the recipe says 'just add lime', where do we get it? In the manufacture of cement, lime is extracted from limestone by heating it to very high temperatures. In the chemical reaction (called calcination), carbon dioxide is released, which seems to make our problem worse, except for the fact that when lime is added to seawater the chemical reaction absorbs almost twice as much carbon dioxide as was released during lime-making. The copious amount of heat needed to extract lime from limestone (cement manufacturing accounts for around 3.4 per cent of all fossil emissions[26]) is usually from burning natural gas, which of course generates carbon dioxide. It is proposed to overcome this problem by capturing the carbon dioxide from burning natural gas and storing it in underground sites. Here the proposal runs into trouble. If we had a well-developed and economic technology for capturing and sequestering carbon dioxide, why would we not simply use it to capture emissions from coal-fired power plants, instead of deploying it to reduce the emissions of a large new industry built to offset one of the effects of emissions from coal-fired power plants?

Instead of natural gas, low- or zero-emissions sources of energy could be used to generate the heat needed to calcine limestone. Suggestions include solar energy, geothermal heat, biomass burning and nuclear power. But if we had a surplus of energy sources to make lime to offset carbon emissions from burning coal, why would we not simply use the surplus energy to replace coal and avoid the problem in the first place? We would – unless the energy supply was not available to substitute for coal because it was so far away from the electricity grid that it would not be commercial to exploit it.[27] So the scheme makes sense only if enough 'stranded energy' can be found. The term suggests the potential energy sources are helplessly isolated. The favoured scheme – promoted by a group called Cquestrate – is to create a vast lime-making facility in the far-flung Nullarbor Plain of southern Australia where there are copious supplies of limestone and abundant unused solar resources.[28] A further advantage is that the Nullarbor seems to have geological strata that lend themselves to burying the carbon dioxide given off during lime-making. The supporters of the scheme estimate that the Nullarbor is big enough to offset the world's annual carbon emissions each year. A quarry 100 metres deep and measuring 10 kilometres by 10 kilometres would need to be dug every year to supply the limestone,[29] although in practice lime-making facilities would be dotted around the globe because the lime would need to be spread throughout the world's oceans. This prompts us to wonder about what it would take to spread the lime around the world's oceans. The answer is suggested by a closely related scheme.

Instead of dispersing lime on the oceans, a related proposal, described by Canadian scientist Danny Harvey, would spread crushed limestone.[30] This has the benefit of avoiding the need to build an energy infrastructure to turn limestone into lime, although it would still require an enormous amount of rock to be crushed, itself an energy-intensive process. A flotilla of ships – some 750 big ones and 3,000 smaller ones, a number Harvey says is 'small compared to the total world fleet of 43,325 ships' – would spread powdered limestone over suitable areas of ocean. The powder would sink, at an optimal rate of 100–600 metres a day, dissolving on the way. Subsurface waters, now enriched with carbonate, would be brought to the surface by upwelling currents and disperse, spreading their alkalinity across the oceans so that they could absorb more carbon dioxide from the air.

Geoengineering using the limestone powder method, like the lime method, would require a huge industrial infrastructure comprising new mining and rock-crushing facilities, an extensive new renewable energy supply (dozens of solar, wind or nuclear power plants, for example), rail systems, port facilities and a 10 per cent expansion of the global shipping fleet.[31] It would require a volume of limestone to be crushed five times greater than is currently crushed each year in the United States. And, more dispiritingly, because deep ocean currents take a very long time to upwell, adding powdered limestone to the oceans would take many decades before the benefits accrued. (In contrast, the addition of lime would take effect within a year.) If 4 billion tonnes were applied each year to the world's oceans, beginning in 2020, and continued for several hundred years, then eventually perhaps an additional 1 billion tonnes of carbon dioxide each year could be taken up by the oceans, reducing the concentration of carbon dioxide in the atmosphere by around 30 ppm, but not until 2200! It might reduce it by twice that amount if we were willing to wait four centuries, until 2500. And that is assuming the world also embarked on a serious effort to reduce greenhouse gas emissions. Another study is less 'optimistic', estimating that adding limestone powder could, by 2500, reduce the concentration of carbon dioxide in the atmosphere by only 17 ppm rather than 30 ppm.[32] Obviously, adding limestone powder to the oceans cannot be regarded as an emergency response, as is true for all carbon dioxide removal methods.

Although not aimed at modifying ocean ecosystems, we would expect adding lime or limestone powder to have ecological effects, especially before it dissolved in deeper layers.[33] One concern is that zooplankton will mistake limestone particles falling through the water for food, which would not be good for their health (nor, of course, is an acidifying ocean). When excreted they would sink more quickly and dissolve more slowly. The extent of this effect is unknown as no tests have been done; surprises are almost inevitable.

One keen advocate of liming the oceans, Greg Rau of the Carbon Management Program at Lawrence Livermore National Laboratory, is enamoured with the 'vast mitigation potential of the ocean'. He argues that it should 'in principle be . . . safe and beneficial' to transform the chemical composition of the world's oceans by adding mineral hydroxide or bicarbonate. He is not fazed by the fact that for every tonne of carbon dioxide removed from the atmosphere at least 7 tonnes of minerals would need to be mined and processed. It would also require global energy generation to expand by a sixth to a half.[34] The cost of developing 'stranded renewable energy' feeds into the preliminary estimate of total cost of US$74 per tonne of carbon dioxide extracted from the atmosphere. With the difference between coal-fired electricity and energy from wind and some kinds of solar power less than US$74, it's not clear why liming the oceans, with all of the uncertainties and spillover effects, would be preferable to cutting emissions by building wind farms and solar plants.

If such renewable energy plants were to utilize stranded energy then the economic question is whether it is cheaper to build long-distance transmission lines for the 'stranded energy' to be linked to the electricity grid (and so substitute for coal-fired electricity) or to keep the coal-fired power plants emitting carbon dioxide and meet the costs of mining, crushing and calcining limestone, transporting lime overland to ports, then shipping it out to sea to be spread over wide areas so that the carbon dioxide emitted from the power plants could be removed from the atmosphere over several decades. In the case of crushed limestone, the cost of calcination would be saved but the benefits would not accrue for several hundred years. It doesn't seem like a difficult decision to make, especially now that climate change is already upon us.[35]

Enhancing weathering

Liming the oceans is related to another geoengineering proposal known as enhanced weathering. Over millennia, rocks break down through contact with rain, which is weakly acidic because of the carbon dioxide it contains. This chemical weathering process forms carbonates (such as calcium carbonate) and the resulting alkaline solution eventually washes into the sea. Climate engineering via enhanced weathering would aim to hasten the natural process. Rocks would be crushed and chemically transformed so that the carbon dioxide gas in the air became embedded in an alkaline bicarbonate solution, which could then be mixed into seawater.

One idea is to take advantage of the fact that carbon dioxide occurs in concentrated form as it flows up the chimneys of coal-fired power plants (around 12–15 per cent compared to 0.04 per cent in the air). Those gases could be pumped through a slurry of crushed mineral carbonate and water. The alkaline bicarbonate solution would then be poured into the sea.[36] According to the proponents, Greg Rau and Ken Caldeira, in a study funded by the Lawrence Livermore National Laboratory, one advantage of this scheme is that, by pumping the alkaline solution into coastal waters, the United States could circumvent international laws such as the London Convention that outlaws dumping chemicals in the open ocean.[37] Although probably at the cost of wrecking coastal ecosystems, in this way the United States could transform the chemical composition of the world's oceans without seeking anyone else's permission.

As with liming, the appeal of the scheme is that oceans could absorb more carbon dioxide from the atmosphere without a noticeable change in their carbon stocks. And it would counter acidification. Yet, as with liming, enhanced weathering would require more than 2 tonnes of rock to be crushed for each tonne of carbon dioxide sequestered.[38] To transport, pulverize and dispose of that quantity of rock would entail an elaborate array of industrial plant, and would generate a huge quantity of waste material. It would also be very expensive.[39] It is another geoengineering scheme that would entail building a vast industrial infrastructure in order to counter the damage done by another vast industrial infrastructure.

Trees, soil and algae

Another suite of geoengineering technologies aims to intervene in land-based biological processes to extract more carbon dioxide from the air. The carbon fixed in trees and other plants as they grow could then be stored, and so taken out of the carbon cycle, or used as a renewable energy source to substitute for fossil fuels, some of which could then be left safely underground. Carbon-sequestering life forms that have attracted the scientific gaze include trees, crops, agricultural wastes, invertebrates in the soil and algae. Larger animals are not much good because they are hard to dispose of permanently. Despite the hype, opportunities for land-based carbon storage as a means of responding to global warming are limited. After all, formed under enormous pressure, fossil fuels store carbon in extremely concentrated form; when above ground carbon is going to occupy a much greater volume.

Nevertheless, we know that in the great carbon cycle some of the carbon in the atmosphere will be fixed in living things as they grow. When they die and decay their carbon atoms are released back into the atmosphere or taken up by other living things in the soil. Although depleted by human disturbance over the centuries, soil carbon remains a large store compared to plant biomass, accounting for some 2,300 billion tonnes of carbon, more than three times the 810 billion tonnes stored in the atmosphere and four times the 650 billion tonnes stored above ground in plant matter, mostly trees.[40] The obvious first approach is to stop clearing forests, because the carbon they store is released into the air when they are burned or decay. (Even if used for building houses or making furniture, the carbon in the wood finds its way into the atmosphere within a few decades.) Deforestation currently accounts for around 11 per cent of annual global greenhouse gas emissions. Of course, the effectiveness of ending deforestation would depend on the carbon emission repercussions of the alternative activities that people undertook to replace the timber and woodchips. Beyond that, schemes to manage rangelands so as to promote vegetation growth and enhance soil carbon have been put forward as means of taking carbon dioxide from the atmosphere and buying time for alternatives to fossil fuels to be adopted. Enhancing soil carbon by better land management is a slow process; so is reforestation, as trees typically take decades to grow. Both are bedevilled by measurement difficulties.[41]

Biological sequestration suffers from an essential flaw – the carbon it stores is in more or less continuous circulation with the atmosphere. As we saw, plants only borrow carbon from the air while they grow, and the debt is repaid when they die. Carbon fixed in soils and trees is a one-off gain (essentially making up for some of the carbon lost from the land in the past) but it is continually threatened with activation and escape into the atmosphere through natural or human disturbance – land clearing, drought, global warming itself and, especially, forest fires. Land-based carbon storage is insecure, unless of course carbon can be buried deeply in fossilized form.

These forms of bio-geoengineering are also limited in their scope. Growing biofuel crops such as sugarcane, corn and forest plantations can substitute for fossil fuels because they too generate energy when burned. But they require substantial inputs – fertilizers, water and, especially, agricultural land that could be used for food production. Similar problems arise with forestry; the land is not available for other uses. If biomass is to be used for energy generation then the facilities are best located near the biomass source or, if the plan is to capture and store the carbon dioxide underground, near the geosequestration site, and ideally near both; otherwise either biomass or carbon dioxide, both bulky, must be transported long distances.

Biochar, which has many enthusiasts, is charcoal created by burning biomass (wood, straw, manure, crop residues and the like) in the absence of oxygen, a process called pyrolysis in which the carbon atoms from the feedstock are bound tightly together and so resist breakdown by microorganisms. Biochar can be added to soils to enhance agricultural productivity by allowing the soil to retain more moisture. Because pyrolysis fixes the carbon, biochar takes it out of the annual flux between biosphere and atmosphere. When we remember that to counter global warming carbon must be sequestered for many centuries, doubts arise about how long the charcoal will hold together before the carbon it stores is released back into the atmosphere. Supporters of biochar claim archaeologists have dug up intact biochar that has been buried for hundreds and even thousands of years.[42] It's not clear whether it is better to convert biomass to biochar (with the concomitant energy use, which would have to be from new renewable sources) or simply to use it as a renewable fuel in power plants to substitute for fossil carbon. In addition, biomass production for biochar may compete with food crops and can itself be energy intensive to produce.[43] Whatever the answers are, at best it seems that biochar has only a small role in offsetting global emissions.

One of the more promising approaches may be to use the carbon-absorbing capacity of fast-growing algae, this time not in the oceans but in ponds. In principle, a hectare of algal aquaculture can yield a much higher volume of biofuel than a hectare of forest plantation or energy crops.[44] However, algae farms require water, which may be scarce, and nutrients, such as phosphate from urea, which has to be manufactured and transported and denied to other uses such as food production. And, as we saw, with phosphate rock in fixed supply and lacking alternative sources, some experts expect 'peak phosphorus' within about 30 years.[45] Algae farming can be carried out on otherwise unproductive land, but to make it carbon-negative the carbon dioxide emissions from using the fuel must be permanently sequestered in exhausted oil and gas fields or other suitable geological formations, or perhaps pumped into the deep ocean. The idea is in its infancy so there are no estimates of the scale of the operation needed to make a difference; but one thing is certain, it would need to be very big.

Carbon dioxide captured from any source (burning fossil fuels, biomass or algae) could be pumped deep underground into geological formations known as saline aquifers. This is known as geosequestration. For it to work, the saline aquifers must be overlain by an impervious rock layer to immobilize the carbon dioxide and prevent it from leaking. As more carbon dioxide is pumped into a formation the pressure becomes intense, and leaks are more likely.[46] In a well-known example, in August 1986 a geological disturbance occurred in Lake Nyos in the Cameroons. Lying in a volcanic crater, the lake is unusual in being saturated with carbon dioxide, which seeps up from magma beneath. One night a cloud of carbon dioxide spilled out of the lake and spread down neighbouring valleys, asphyxiating 1,700 people and 3,500 cattle; in fact every animal and bird within a 25 kilometre radius died.[47] Nevertheless, assuming no leaks from the saline aquifers, eventually the carbon dioxide would be rendered relatively safe by being dissolved into the brine, although the process would take centuries.

There are strong arguments for preferring storage of carbon dioxide in rock sediments under the sea rather than on land.[48] First, the extent of suitable marine sediments (such as permeable sandstone capped with impermeable rock layers) is very large. Second, and counter-intuitively, the brine in marine sediments is much less salty than that in terrestrial saline aquifers. It is essentially filtered ancient seawater. If it leaks into the sea as carbon dioxide is injected, it will do little damage, unlike the ultra-salty and toxic brine of terrestrial formations that might bubble up and poison the landscape.

Carbon capture and storage is not, strictly speaking, a form of geoengineering because it applies to identifiable sources of carbon emissions rather than the regulation of the atmosphere as a whole. Nevertheless it has its strong supporters among the geoengineering community. It is expensive, and will remain so. One expert has calculated that capturing just a quarter of the emissions from the world's coal-fired power plants would require a system of pipeline infrastructure big enough to transport a volume of fluid twice the size of the global crude-oil industry.[49] Once touted as the saviour of the coal industry, enthusiasm for carbon capture and storage has waned as predictions by doubters about high costs and technical difficulties have proved accurate. It is hard to see why it would become worthwhile as an add-on to new energy industries when it has not been added on to old ones. Yet a number of geoengineering schemes propose to build two immense new industrial infrastructures, one to extract carbon from the atmosphere and one to bury it underground, all aimed at molli-fying the owners of an existing industrial infrastructure.

Purifying the air

Instead of relying on trees or algae to soak up carbon dioxide, research is proceeding into how to extract it directly from the air. Using well-known industrial processes, air can be blown across surfaces covered with water and chemicals, such as sodium hydroxide (caustic soda), to generate carbonate solids. The carbon dioxide is then extracted by heating, usually with natural gas. 'It's a big, ugly industrial process', says one of its proponents, 'that uses at almost every step hardware you can buy commercially today.'[50] For decades chemical engineers have been using sodium hydroxide to extract carbon dioxide from the air in submarines and space ships. It may seem strange to want to extract carbon dioxide from the ambient air, where its concentration is less than 0.04 per cent (395 ppm) when it is around 12–15 per cent in the flue gases of coal-fired power plants.

What would be needed to deploy enough direct air capture technology to make a difference to global warming? A technology assessment for the American Physical Society has done some calculations.[51] A typical air capture machine might look like a long metal box, 10 metres high and 1 kilometre in length. To extract 1 million tonnes of carbon dioxide each year, an array of five would be required covering an area of 1 square kilometre, allowing 250 metres between each long box so that there was enough space for the carbon dioxide-depleted air to be replenished. The array would be attached to a chemical plant to separate out the carbon dioxide. Another entire infrastructure would be needed to transport and bury the waste underground. A standard-sized 1,000 megawatt coal-fired power plant emits around 6 million tonnes of carbon dioxide each year, so it would need six of these arrays to offset its annual carbon dioxide emissions. It would need 30 kilometres of air-sucking machinery and six chemical plants, with a footprint of 6 square kilometres.

The material and energy resources required to construct this sprawling industrial infrastructure – including the factories needed to produce the steel and the chemical plants needed to make the sodium hydroxide – would be daunting. Let's say in a few decades, after cutting emissions to zero, the world decides to reduce the concentration of carbon dioxide in the atmosphere by 50 parts per million (say from 500 ppm to 450 ppm, remembering that the pre-industrial level was 285 ppm and it is now 395 ppm). So we start to build 13 of these 1 kilometre square arrays each year. At that rate it would take a century of building activity.[52] Spread out around the world these 1,300 industrial facilities would act like a gigantic air purifier for the planet, giving new meaning to the phrase 'spaceship Earth'.[53]

The idea of extracting carbon pollution from the air prompts an analogy with mercury pollution from chemical factories in the Great Lakes. There are in principle four kinds of response. The first is to change industrial processes to eliminate mercury. In the case of global warming, this is equivalent to switching to renewable energy – the carbon pollution is not generated. The second is to install pollution control equipment on the ends of the pipes, to stop the mercury entering the lakes. This is equivalent to carbon capture and storage. The third is to filter the lake water somehow to extract the mercury, equivalent to carbon dioxide removal technologies like air capture and ocean liming. The fourth is to issue prophylactic tablets that counter the effects of mercury poisoning to those whose water is supplied by the lakes. This symptomatic approach is equivalent to solar radiation management, considered in the next chapter.

It might seem, at first blush, that ideas such as iron fertilization, ocean liming and growing more trees to soak up carbon dioxide are plausible responses to rising levels of carbon dioxide in the atmosphere. As we take a closer look, however, things become much more complicated, and 'on each solution a mystery waits to leap'.[54] Other factors and forces come into play that turn simple ideas into complex interventions, the ramifications of which we understand incompletely or hardly at all. We begin to have the vertiginous feeling that, for all of our wonderful scientific advances, we don't know much about the Earth system at all. And the thought arises that before we began to disturb it there was a pattern and logic to the distribution of carbon atoms in various states around the planet – in the oceans, the biosphere, the rocks, the air and the deep-earth deposits – a pattern linked to the evolution of life itself. Releasing carbon from its subterranean tombs and emitting it into the air, where the amount has risen by nearly half and will soon double, cannot be reversed because there are no other storage places in which we can have confidence.

The essential difficulty with all carbon dioxide removal approaches is that they want to push a reluctant genie back into the bottle. It took the Earth millions of years to immobilize a large portion of the planet's carbon in fossilized form deep underground. When we extract and burn it we mobilize the carbon and there is no place on Earth where, over human timescales, we can safely sequester it again. We know we cannot leave it in the atmosphere. Carbon stored in vegetation and soils is always on the brink of release through fire or human disturbance. The oceans are in constant flux, with even the deepest layers naturally coming to the surface sooner or later. Heavily promoted plans for carbon capture and storage, in which carbon dioxide extracted from the smokestacks of coal-fired power plants is pumped into geological repositories underground, looks increasingly risky and expensive (as we will see in chapter 7). I hope we have learned enough by now to be wary of any technology that claims to have found a way to immobilize for centuries huge quantities of carbon somewhere in the Earth system where it does not belong. Even if such a place could be found there is something deeply perverse in the demand that we construct an immense industrial infrastructure in order to deal with the carbon emissions from another immense industrial infrastructure, when we could just stop burning fossil fuels.

3

Regulating Sunlight

Solar radiation management

Human-caused changes to the atmosphere have upset the natural energy balance of the Earth. The enhanced greenhouse effect warms the globe because higher concentrations of carbon dioxide (along with other greenhouse gases like methane and nitrous oxide) cause the atmosphere to absorb more heat near the Earth's surface. In technical terms, the Earth always seeks to balance the energy it absorbs from sunlight with the energy it emits as infrared radiation back to space (the so-called radiative budget – energy in equals energy out). When more energy is trapped in the Earth system it heats up so that, at the warmer temperature, energy going out will once again balance with energy coming in.

Where is the extra energy being stored? In the shorter term the atmosphere warms up, but in the longer term the heat energy is stored in the oceans. With their higher volume and density, the heat-storing capacity of ocean waters dwarfs that of the atmosphere. The amount of energy required to warm the planet's atmosphere by 1°C would warm only the top 3 metres of the oceans by the same amount.[1]

Solar radiation management technologies are designed to regulate the energy balance by reflecting a greater proportion of sunlight back to space. Unlike carbon dioxide removal methods, which aspire to control the carbon cycle, solar radiation management aims to manipulate the primary source of energy that makes the Earth a living planet. It attacks a symptom of the disease, a warming globe, rather than its source, rising greenhouse gas emissions, and leaves other symptoms, notably acidifying oceans, untouched. Below I assess three proposed solar radiation control technologies – marine cloud brightening, cirrus cloud modification and, the big one, sulphate aerosol spraying. I devote much less space to the first two because they receive much less attention from researchers and, at this point, seem much less likely to be implemented than the third.

Brightening clouds

Not all of the Sun's light reaches the Earth's surface; some is reflected back into space by clouds, and the whiter they are the more they reflect.[2] One idea is to cool the Earth by modifying the low-lying sheets of stratocumulus cloud that cover around a quarter to a third of the world's oceans so that they become more reflective. Success depends on finding a mechanism for influencing the albedo of these marine clouds, that is, their reflectivity. The answer is provided by the 'Twomey effect' according to which the albedo of low-lying clouds rises with an increase in the number of cloud condensation nuclei, tiny particles or aerosols that act as seeds around which water vapour condenses. Clouds that contain a larger number of small droplets have a greater surface area to reflect sunlight. Many types of tiny aerosol particles can be used artificially to promote cloud condensation, the classic being silver iodide, used for many years for cloud seeding. But dust, soot, volcanic sulphates and sulphate aerosols from oxidation of dimethyl sulphide produced by phytoplankton serve the purpose. So does sea salt.

The idea behind marine cloud brightening – championed by John Latham, a professor of atmospheric physics at the University of Manchester, and Stephen Salter, an engineering professor at the University of Edinburgh – is to construct a fleet of special ships that would roam the oceans pumping submicron-sized drops of seawater into the air. The drops would need to be pumped only around 30 metres, after which air turbulence would in the right conditions mix them vertically through the planetary 'boundary layer', the top of which can be some 300 to 2,000 metres above sea level.[3] There the spray would evaporate, leaving salty residues that would promote condensation in stratocumulus clouds. Infused with a larger number of smaller drops, the clouds would reflect more solar radiation back into space. Within days or weeks the particles are washed out by rain and so to sustain the effect spraying would need to be continuous.

Not all areas of ocean are suitable: the air above needs to be sufficiently moist and have the right kind of air turbulence. It's estimated that increasing average cloud reflectivity from the usual 50 per cent to 60 per cent would reflect enough solar radiation to offset the warming effect of a doubling of carbon dioxide concentrations.[4] If we reached that point some 1,500 unmanned, satellite-controlled vessels would be needed to patrol the sea. Fewer vessels would be needed if deployment began sooner, but the number would grow each year that emissions were not reduced. The tiny size of the sea-spray droplets would require each vessel to be fitted with around 28 billion (yes, billion) nozzles of diameter a little less than one micron (millionth of a metre). The nozzles would have be etched chemically into silicon wafers using a process called micro-fabrication lithography.[5]

One advantage of the proposal is that it does not entail introducing foreign substances into the environment; but, as we would expect, brightening marine clouds will have a chain of effects beyond reflecting more sunlight back into space. In regions where clouds are modified to increase their reflectivity, the ocean would be cooler, disturbing ocean circulation patterns, which in turn may alter precipitation patterns on land, depending on, among other things, the season.[6] The models that try to simulate these linkages are inadequate and more work is needed. A variation on this idea has already found its way into popular culture. The plot of Clive Cussler's thriller *The Storm* centres on an Arabic Dr No who controls a technology that can manipulate ocean currents so as to shift the monsoon. He plots to move the rains from India to arid regions in north Africa and southern Eurasia, where his wealthy associates have bought up land.[7] Rather than manipulating marine clouds, the technology deploys billions of nano-robots that devour all organic matter in the Indian Ocean, though exactly how this allows the monsoon to be directed is left to the reader's imagination.

Almost nothing is known about how long-term spraying of ocean tracts would interact with global climate patterns such as continental warming, ice-sheet melt and methane release. However, some early work indicates that brightening clouds in one region may cause marked climatic changes in remote regions, perhaps because there is global competition for water vapour. The three best sites for marine cloud brightening are in the North Pacific (off the north-west coast of the United States), the South Pacific (off the coast of Chile) and the South Atlantic (off the west coast of southern Africa). Models indicate that brightening clouds in these regions can bring baffling changes to the weather on the other side of the world.[8] For example, when cloud brightening is carried out in the North Pacific more rain is expected in South Australia. When clouds are brightened in the North and South Pacific, rainfall is predicted to increase in the Amazon. But spraying in the South Atlantic has the opposite effect, a decline in Amazonian precipitation.[9] When intervention occurs at all three sites, the Atlantic effect seems to dominate the Pacific effect. All of this is puzzling, to say the least, and serves as a stark reminder that we are a long way from knowing enough to be confident that interfering with the climate system will not have dire unintended effects. John Latham, who developed the idea for brightening marine clouds, has said that if the Amazon drought effect is confirmed it would be the death of the scheme.[10]

One study indicates that enhancing the Earth's albedo by marine cloud brightening would seem to work much more effectively when seeding occurs in clouds formed in pristine air masses.[11] Seeding is only around a quarter as effective when carried out in a 'dirty air mass', one polluted by industrial emissions or by ship exhausts. Ships burn low-grade diesel and are notoriously polluting. The albedo response also depends on meteorological conditions; there are some regions where artificial aerosol plumes actually cause the water content and thus the albedo of marine clouds to fall.[12] In that case, the intervention would make global warming worse.[13] An apparent advantage of a programme of marine cloud brightening is that it would be easy to terminate; within a week or two the skies would return to 'normal'. As we will see, however, the sudden termination of any solar radiation technology could be disastrous because the heating suppressed by the intervention would rebound at a much faster rate. Moreover, a system that can be turned off quickly for the right reason can also be turned off quickly for the wrong reason, such as a government responding to an ill-founded panic.

Modifying cirrus

Cirrus clouds are long, wispy clouds that form at altitudes above 6 kilometres. They typically shade around 45 per cent of the Earth and are more common in the tropics.[14] Although not well understood, they are thought to be pivotal in climate change because they both reflect incoming solar radiation and absorb outgoing thermal radiation.[15] But they prevent more heat from escaping than they allow in, so their net effect is to warm the planet. If we could eliminate cirrus clouds the Earth would be cooler, a thought that led to another geoengineering idea.[16] Unlike cloud brightening, which seeks to reflect solar radiation before it reaches the Earth's surface, removing cirrus clouds endeavours to clear the way for more heat emitted from the Earth to escape to space. David Mitchell, the scientist who dreamt up the idea, suggests that this method of geoengineering is perhaps better described as thermal radiation management than solar radiation management.[17]

The best way to reduce cirrus clouds, it is thought, is to change the process by which ice crystals are produced in them (called nucleation), which may also hasten the rate at which ice in them aggregates into larger crystals (called aggregation). Larger ice crystals fall out faster, reducing the global coverage of cirrus clouds. We might be able to accelerate this process by injecting a non-toxic chemical known as bismuth tri-iodide into the hot exhaust gases of commercial aircraft, which routinely fly through regions with cirrus clouds. Bismuth is a non-toxic metallic element used in cosmetics, fireworks and pharmaceuticals. It lends its name to Pepto-Bismol, taken since 1901 for upset stomachs. (Pepto-Bismol is administered to seabirds contaminated by oil spills to flush oil from their intestinal tracts.) The chemical (which is much cheaper than silver iodide) would infuse the atmosphere with bismuth tri-iodide aerosols, producing larger ice crystals in cirrus clouds. This chemically modified atmosphere would, if the plan worked, allow more solar radiation to reach the Earth's surface but allow a greater amount of this energy to leave as heat radiation. It might have the added benefit of turning air travel into an environmentally beneficial activity; instead of carbon offsets, perhaps airlines could offer bismuth credits.[18] Cirrus clouds in the mid-latitudes and near the poles may be the best target because the scheme appears to work better on cirrus clouds unrelated to thunderstorms (and climate sensitivity is greater near the poles). Like all solar radiation management technologies, cirrus cloud modification would do nothing to halt the acidification of the oceans and its wider effects on the global climate are unknown. As a general principal it is safe to assume that cirrus clouds are there for a reason and that taking them away would have complex ramifications.

Spraying sulphur

The idea of cooling the planet by spraying sulphate particles into the upper atmosphere was sparked by observing the effects on the weather of volcanic eruptions. American polymath Benjamin Franklin attributed the abnormally cold winter of 1783–4 to the 'dry fog' that had for months enshrouded the northern hemisphere following a huge volcanic eruption in Iceland. The particles injected into the stratosphere by the eruption of Mount Laki, which began in June 1783 and lasted with declining intensity for eight months, were responsible for one of the most severe winters on record, reducing the average northern hemisphere temperature by 1.3°C, and by as much as 3°C in central England.[19] Franklin, living in Paris at the time, noted:

This fog was of a permanent nature; it was dry, and the rays of the sun seemed to have little effect towards dissipating it ... They were indeed rendered so faint in passing through it, that when collected in the focus of a burning glass, they would scarce kindle brown paper.[20]

In the French capital that winter, firewood was soon unattainable. Around the world, the weather was disrupted. In Japan, famine followed the failure of the rice crop.

The year 1816 became known as 'the year without a summer'. The cause of the missing season was Mount Tambora in Sumbawa, Indonesia. Its eruption some months earlier is classified as 'super-colossal', much bigger than any other in recent centuries including Laki and Krakatoa in 1883. The aerosol veil that enveloped the world brought on a cold, wet winter in the United States and Western Europe. Food riots broke out in England. In summer the spectacular sunsets caused by the haze found artistic expression in the red skies of William Turner's paintings. Holidaying in the Swiss Alps, 18-year-old Mary Godwin was trapped indoors by 'incessant rainfall'. To pass the time she and her companions, Percy Bysshe Shelley and Lord Byron, challenged each other to concoct horror stories. Inspired by a dream, Mary told a tale that would three years later become the novel *Frankenstein, or the Modern Prometheus*. Byron drafted 'Darkness', a poem in which 'the bright sun was extinguished'.

So at least since Mount Laki exploded, it has been known that large volcanic eruptions change the weather. In 1991 the ash poured into the atmosphere by the eruption of Mount Pinatubo dimmed the Earth enough to cool it by around 0.5°C for a year, returning to normal over the next two years as the ash cloud fell out of the air.[21] These 'natural experiments' prompted some climate scientists to conceive the idea of countering global warming by mimicking the cooling effect of volcanoes. Stratospheric aerosol spraying is the archetypal geoengineering technique – it would be easy, effective and cheap, and have the most far-reaching implications for life on Earth.

The stratosphere is the layer of the Earth's atmosphere stretching from about 10 to around 50 kilometres above the surface (although higher in the tropics); below it is the troposphere where all of the weather occurs. For perspective, the peak of Mount Everest is almost 9 kilometres above sea level and commercial airliners cruise at altitudes of 9–12 kilometres. Compared to the troposphere, where they rise and fall, air masses in the stratosphere flow horizontally, so particles remain in the higher layer for much longer, one to two years. In the troposphere they typically last only one or two weeks before falling out or being washed out by rain. To complete the sketch, it's worth noting that in the troposphere as altitude increases temperature falls, while in the stratosphere the reverse applies – it's colder at the bottom than at the top. The inflexion from cooling to warming occurs in a narrow layer between the two called the tropopause.

The proposal is to spray tiny aerosol particles into the stratosphere in order to reflect an extra 2 per cent or so of incoming solar radiation, about what it would take to offset the global warming associated with a doubling of greenhouse gases in the atmosphere. Although aluminium-based particles, soot and customized nanoparticles have been suggested, most work focuses on sulphur, which would be sprayed in the form of sulphur dioxide, hydrogen sulphide or sulphuric acid. Each acts as a 'precursor' that quickly combines with dust and water to make sulphate aerosols. The most likely delivery method is a fleet of customized high-flying aircraft fitted with tanks and spraying equipment, although naval guns, balloons and a hose suspended in the sky have also been suggested. In effect, humans would be installing a radiative shield between the Earth and the Sun, one that could be adjusted by those who control it to regulate the temperature of the planet.

How much sulphur would be needed to block around 2 per cent of incoming solar radiation? Paul Crutzen estimated that around 5 million tonnes would be needed annually, although others suggest it might be somewhat less if the particles are smaller.[22] Smaller particles have a proportionally larger surface area and so reflect more light. Five million tonnes is around a tenth of the amount of sulphur pollution emitted to the lower atmosphere in 2005 from fossil fuel combustion and industrial processes.[23] As we saw, for the same cooling effect much less is needed in the upper atmosphere because each particle stays aloft perhaps 50 times longer and so does 50 times more work. The comparatively small quantities mean that when sulphate aerosols fall out of the stratosphere they will not add appreciably to acid rain. The greater diffusion of light would result in whiter daytime skies and redder sunsets. (The spectacular sunsets after Krakatoa inspired Edvard Munch in Oslo to paint *The Scream*.)[24] Some plants prefer diffuse rather than direct sunlight; other things being equal, they would grow more quickly. While the acid rain effect would not be great, the effect of aerosol spraying on bio-geochemical processes on land and in the oceans is complex, a fact only now beginning to be recognized.[25]

The eruption of Mount Pinatubo is estimated to have injected around 10 million tonnes of sulphur (as sulphur dioxide) into the stratosphere. Regulating the Earth's temperature with a solar filter would be equivalent to one Mount Pinatubo every four years.[26] In their evaluation, one group of scientists estimates that if the aerosol were delivered by a fleet of aircraft the size of jet fighters then a million flights a year, each of four hours duration, would be needed.[27] The programme would require a fleet of several thousand aircraft. If such a fleet were available, or a hose could be persuaded to stay aloft, one of the virtues of sulphate aerosol spraying is that it requires no major technological innovation, only some engineering, so a programme could be implemented quickly. Moreover, once begun it would take effect within a few months. Economists are impressed by the fact that it would be cheap, a small fraction of the estimated cost of cutting carbon emissions.[28] This has its drawbacks, as we will see, for it removes a major hurdle to any nation, or even a rogue billionaire, who might be tempted to tinker with the climate.

The effectiveness of the solar filter would depend on the emission strategy adopted. In addition to the amount of sulphur injected, a number of variables would have a bearing on the amount of cooling.[29] For this reason the volcano evidence, while suggestive, is in fact not very helpful; after all, large eruptions have been associated with unseasonal cold, drought, famine and food riots. The cooling they bring also changes ocean currents, which can prolong their climate impacts for 20–25 years after the eruption.[30] The particles from an eruption are twice the optimal size, and the cooling effect is much stronger at the poles than in the tropics. A volcano affects the climate for a few years at most; modifying the planet's atmospheric chemistry over very long periods is another kettle of fish entirely. There are various other complicating factors. Doubling the amount of sulphur injected will not have twice the cooling effect. Location matters because, for example, dispersal is more effective if done in the tropics. And choosing the altitude of the injections would need to balance the trade-off between higher injections that reflect more sunlight, and greater damage to the ozone layer. In addition, studies indicate that a programme of continuous injections would not have the same effect as the same amount injected in a few pulses each year.[31] Aerosols formed after spraying sulphuric acid are smaller and have a longer lifetime than those generated from spraying sulphur dioxide.[32]

Any geoengineering relying on sulphate aerosol spraying will need an emission strategy that takes account of all of these factors, yet our understanding of them is rudimentary, to say the least.

How effective would a solar filter be in suppressing warming? The most commonly cited study, by Ken Caldeira and Lowell Wood, used a fairly simple climate model to simulate the effects on warming around the world.[33] The model generates a map showing a warmed-up world with extreme temperature rises towards the poles when carbon dioxide concentrations are doubled. But when solar radiation reaching the Earth is reduced by 1.84 per cent, the world's temperature pattern returns almost to normal, although still with some warming near the poles. Thus the Royal Society rates sulphate aerosol spraying highly effective at countering warming.[34]

In the case of rainfall, Caldeira and Wood claim that the solar shield is quite effective at restoring normal patterns around the world. However, a more recent and comprehensive study by a European team suggests a quite different set of impacts.[35] They begin by making the disconcerting assumption that carbon dioxide concentrations quadruple. Bearing in mind that we have been expending nervous energy on the prospect of a doubling of carbon dioxide, this comes as a shock because a fourfold increase is associated with temperatures some 8°C higher over land surfaces.[36] It would be 'game over' before we reached that level. Nevertheless, the researchers point out that an increase from the pre-industrial level of 280 ppm to some 1,100 ppm is towards the top of the range of estimates for the end of the century. Besides, simulating such a large change in carbon dioxide concentrations allows a clearer picture of the effectiveness of a solar filter.

They calculate that the amount of sunlight reaching the planet would need to be turned down by around 4 per cent to counter the warming effects of a quadrupling.[37] One problem is that carbon dioxide spreads quickly around the planet, trapping more heat wherever it goes, while schemes to deflect light from the Sun would have a greater effect over the tropics where solar radiation is more intense. So if a solar shield returned the Earth to a pre-industrial temperature on average, the tropics would be cooler by about half a degree while the poles would be warmer by 1–2°C, enough to melt a lot of ice.

But it is the effect on rainfall that raises most concern. While rainfall is expected to increase with a warming globe, reducing solar radiation enough to force temperatures back down would weaken the global hydrological cycle, meaning less precipitation than in the pre-industrial climate. In the tropics and parts of northern Europe and North America rainfall is projected to decline by 10–20 per cent below pre-industrial averages, mainly in summer. Rainfall in parts of the Amazon is also expected to fall by around 20 per cent.[38] Taking account of changes in evaporation as well as precipitation does not appreciably alter the picture.

Other experts argue that the complexity and opacity of the climate system mean that it is impossible to draw any such conclusions about what would actually happen if we tried to adjust solar radiation to cool the planet. The chemistry of the atmosphere is complicated, so simply turning down the amount of sunlight reaching the Earth, as the models do, can give little clue as to what would happen in the real climate system. Particles sprayed into the stratosphere are subject to a range of microphysical processes, including nucleation, condensation, evaporation, coagulation, sedimentation and washout (in the troposphere).[39] This is not the place to describe each of these; but to give a flavour, one study concludes that after injection the particles will grow because they clump together (coagulation) or merge with other aerosols (condensation). Bigger particles reflect less solar radiation, but they also sediment, that is, they fall because gravity reduces their residence time and thus the reflective capacity of a given amount of sulphur dioxide. So studies that do not take full account of atmospheric chemistry seriously underestimate the amount of sulphur that would need to be pumped into the stratosphere.[40]

Confidence among geoengineering scientists in the efficacy of solar radiation management was shaken by a study showing that the Indian monsoon could be seriously disrupted, affecting food supplies for up to 2 billion people,[41] although the disruption might be less than in a scenario of warming without the solar filter. A later study concluded that the effects on the Indian monsoon might be mixed, with no decline over India but a weakening over Southeast Asia.[42] Perhaps it's worth noting here that over the second half of the twentieth century South Asia experienced a drying trend in summer, with annual summer rain falling by around 5 per cent. The decline in rainfall is due principally to more aerosols in the atmosphere (mainly associated with burning fossil fuels) which, via dimming, have slowed down the tropical weather pattern responsible for the annual monsoon. Against this pollution-induced trend, a warming world holds more moisture in the atmosphere, which leads to projections that monsoon rainfall will increase.[43] The implication is that the aerosol haze over South Asia has masked an increase in precipitation due to rising greenhouse gases.

Nevertheless, our understanding of what influences the monsoon is weak, our knowledge of how global warming would change the monsoon is weaker, and trying to estimate the combined influence of warming and solar radiation management (along with anti-pollution measures) is little more than educated guesswork. Who knows what would happen to rainfall, but if catastrophe ensued after sulphate spraying at least we would know whom to blame. Or would we?

High in the stratosphere lies the ozone layer, which serves the vital role of protecting living things from the intense flux of ultra-violet radiation streaming from the Sun. The Montreal Protocol, in force since 1989, mandates the phasing out of ozone-depleting chemicals and it is expected that the hole in the ozone layer over the Antarctic will be repaired by 2050. So one important question is whether the extra sulphur compounds put into the stratosphere would interact with ozone. The most comprehensive study concluded that injecting enough sulphur to suppress the warming associated with a doubling of carbon dioxide concentration would indeed deplete ozone in polar regions, delaying the recovery of the Antarctic ozone hole by 30–70 years.[44] This has led some Russian researchers to argue that sulphate aerosol spraying should not be considered until the ozone hole has been repaired.[45] Although an admirably cautious approach, it could also be framed as 'now that we have healed the ozone layer, we are free to harm it again'.

A further anxiety about dimming the globe with sulphate aerosols, perhaps the greatest once it is grasped, is the danger of bringing the programme to a sudden end, a move that might be deemed necessary by political turmoil or international conflict, or the realization that one of its side effects (such as monsoon failure or ozone depletion) is much worse than anticipated. As aerosol spraying suppresses warming, but does not eliminate its cause, an abrupt end would result in a sudden leap in global temperatures. This is known (without irony) as the termination problem and is illustrated dramatically in figure 3, in one of those charts that climate science occasionally produces that sends a chill down the

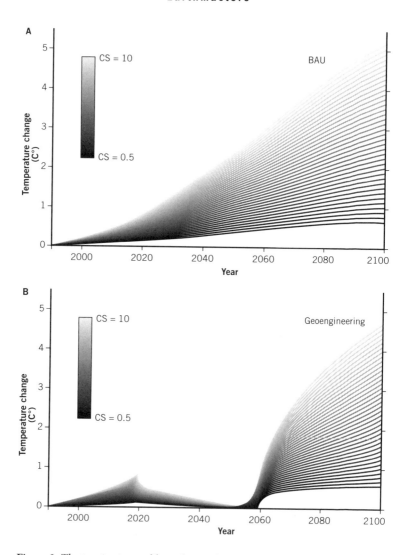

Figure 3 The termination problem: chart A shows temperature change compared to the year 1990 for the business as usual (BAU) scenario, with different shades representing different assumptions about the sensitivity of the climate (CS) to increased concentrations of greenhouse gases in the atmosphere; chart B shows temperatures when geoengineering is applied from 2020 to 2059, then halted. (Note that the best estimate of climate sensitivity is 3°C, that is, average global warming of 3°C for a doubling of the concentration of CO_2 in the atmosphere.)

spine. Climate modellers Andrew Ross and Damon Matthews have estimated the likely path of warming if a programme begun in 2020 were to be suddenly stopped in 2059.[46] Under the most likely scenario, the termination of aerosol spraying in 2059 would see a surge in average temperature by 1.3°C in the first decade, falling back to 0.33°C in the following decade.[47] For some time ecologists have stressed that the *rate* at which the globe warms is a greater threat to ecosystems than the amount of warming. Slower warming gives plant and animal communities more time to adapt. One study concluded that if warming occurs at a rate of 0.1°C per decade, half of ecosystems can adapt. The other half cannot. At a warming rate of 0.3°C per decade, only 30 per cent of ecosystems can adapt, and among forest ecosystems the measure shrinks to 17 per cent.[48]

So, few ecosystems would survive the precipitate rise in temperature should the solar filter be turned off. Large parts of the globe would be denuded. After the first disastrous decade, warming of 0.33°C per decade remains extremely high, and compares to the expected warming under the business-as-usual scenario, that is, warming without geoengineering, of a little under 0.3°C, itself a rate of warming too fast for all but a sixth of the world's forests to survive. Once deployed it is likely that we would become dependent on our solar filter, the more so if we failed to take the opportunity while it was in place to cut greenhouse gas emissions sharply. This is perhaps the solar filter's most dangerous drawback.

For all of the dramatic evidence provided by Vulcan's 'natural experiments', it has been argued that there is one 'killer objection' to global dimming via sulphate aerosol spraying – it cannot be tested without full-scale implementation.[49] A testing programme that sprayed a thousand tonnes of sulphur dioxide into the stratosphere would tell us virtually nothing about the impact on the Earth's climate of full deployment. Moreover, if it were fully implemented, ten years of climate data from around the world would be needed in order to separate out the effects of the solar filter from other causes of climate variability. If after a couple of years climatic disasters occurred – droughts in India, for example – we would not know whether they were caused by global warming, the solar filter or natural variability. India would demand that the programme be suspended; the experts at a future World Climate Regulation Agency would say the cause wasn't known and turning off the solar filter could make matters worse. Despite all this, the Royal Society assesses sulphate aerosol spraying as 'the most promising' of solar radiation management methods.[50] And in a comprehensive review, a team of scientists drew the following less than reassuring conclusion: 'Observations following major volcanic eruptions have demonstrated that sulphate aerosol, in sufficient amounts, will cool the planet, and that the Earth system can survive this kind of perturbation.'[51] 'Survival' sets a pretty low bar.

Tailoring the solar filter

Enshrouding the globe in a haze of sulphur pollution seems like a blunt and indiscriminate method, so some scientists have begun work on refinements. David Keith has suggested the creation of a new kind of nanoparticle that could lift itself into the stratosphere by exploiting a phenomenon known as 'photophoretic levitation', a complex process in which temperature differences in airborne particles can cause them to rise.[52] It would allow more control, he claims, and might be less harmful to the ozone layer. It is apparent, I think, that injecting exotic particles into the stratosphere, with its

complex chemistry, would be a gamble, not least because full-scale engineering would be necessary before the impact could be reasonably assessed. 'Engineered particles', write some sceptical researchers, 'may have unknown and unforeseen effects, and their residence time in the atmosphere will be unknown until full-scale atmospheric experiments are conducted.'[53] Residence time matters; half the lifetime requires twice the amount, yet long residence makes it slower to turn off the experiment if things go awry.

Ken Caldeira and Lowell Wood have argued that the implementation of climate engineering could be tailored to obtain the desired mix of temperature and rainfall. The Arctic, prone like the Antarctic to much higher warming than lower latitudes, is especially attractive for manipulation. The size of the Arctic ice cap could, they suggest, 'be arbitrarily adjusted by varying insolation to various extents over different areas'.[54] You want ice? Just say how much. Their model shows that instead of summer sea-ice declining by 44 per cent below its pre-industrial extent with a doubling of carbon dioxide, they can expand it by 136 per cent above normal by turning down the sunlight over the North Pole by 12 per cent.[55] Man imitates God.

Mike MacCracken, formerly head of atmospheric sciences at Lawrence Livermore National Laboratory, is the chief scientist at the Climate Institute, a Washington-based environmental organization. He has been engaged in geoengineering research since his time at Livermore.[56] He is particularly anxious about the possibility of a climate emergency within the next decades but takes the view that we should not wait for it to occur but deploy a solar filter as soon as possible.[57] Instead of attempting to control the entire climate system MacCracken has identified a number of 'high priority applications' that would be targeted at particularly

worrying manifestations of global warming. They include injecting sulphate aerosols over the Arctic to reduce warming and promote the build-up of mountain glaciers and ice sheets in the region, supplemented with marine cloud brightening to reduce warming in patches of ocean so as to reduce the intensity of tropical cyclones. He also suggests that the warming that will follow the clean-up of urban air pollution in populated regions of China and India be offset by injecting sulphate aerosols over an appropriately sized area of ocean in the tropics. So instead of the sulphate aerosol 'umbrella' hanging low over cities it would float high above the oceans, with the sulphur injected perhaps from mountain tops or from elevated hoses anchored on Pacific Islands.[58]

We find ourselves in an exquisite dilemma. Sulphate pollution from burning coal and oil has a cooling effect on the planet yet the thick brown haze covering much of Asia and other conurbations is estimated by the World Health Organization to kill 1.3 million people each year.[59] The sulphates in this lower-atmosphere pollution have been so effective at offsetting global warming that without it, on top of the measured 0.8°C warming since pre-industrial times, the Earth would be an extra 1.1°C warmer.[60] As the governments of China, India and other industrializing countries follow the example of Western nations and introduce air pollution laws to improve public health, the latent warming will become manifest.[61] The lifetime of sulphate aerosols in the lower atmosphere is one or two weeks while the molecule it is meant to counter, carbon dioxide, stays up there for many centuries. So if we were to stop burning fossil fuels tomorrow, and eliminate carbon dioxide emissions, the planet would immediately become warmer, and remain so for some decades. It would be the equivalent of the concentration of carbon dioxide in the atmosphere leaping from 390 ppm to 490 ppm within a few weeks.[62]

This is a truly frightening fact. If world leaders were persuaded to agree to a programme of rapid reductions in carbon emissions, we might need somehow to maintain levels of sulphur pollution in order to avoid a warming so rapid that many ecosystems could not survive. The only answer seems to be to maintain this level of pollution for many decades until enough carbon dioxide can be shifted out of the atmosphere by natural or artificial means.

4

The Players and the Public

The geoclique

Although still in its early days, the constituency for geoengineering is now developing around a network of individuals with personal, institutional and financial links. At the centre of the network is a pair of North American scientists actively engaged in geoengineering research – David Keith and Ken Caldeira. Keith and Caldeira have been dominant voices in virtually every inquiry into or report on geoengineering.[1] They have been prominent expert witnesses at the opinion-forming House of Commons inquiry in the United Kingdom and the technology assessment of geoengineering carried out by US Congress's Government Accountability Office.[2] And their views have helped frame the deliberations of the Intergovernmental Panel on Climate Change as it grapples for the first time with the scientific and ethical tangle that is climate engineering.

In short, Keith and Caldeira are ubiquitous and have become the 'go to' guys on climate engineering. Such is their influence over the global debate that author Eli Kintisch has dubbed them the 'geoclique'.[3] While they are not as gung-ho as some other geoengineering advocates, their research and advocacy lead in only one direction. The course of events is on their side. The urge to mastery

over nature is inscribed in the climate engineering project, and it is the momentum of this urge that will overwhelm the best efforts of the reluctant geoengineers.

David Keith is a physicist, entrepreneur and professor of public policy. For many years he was based at the University of Calgary before moving to Harvard. Although there is no doubting his brilliance as a scientist, his views are sometimes hard to pin down. While convinced of the validity of climate science, he seems to adopt a nonchalant stance towards its impacts on humans. 'I'm not sure that global warming is such a threat to human civilization . . . human beings are a remarkably adaptable species. . . . If it is just the human race you're worried about, I'm not sure global warming is such a big problem.'[4] He expects that humans will be engaged in 'planetary management' via climate engineering and what remains of the natural world will be managed like a garden, a development he seems to accept with equanimity.[5] Keith is pushing ahead with plans to test sulphate aerosol spraying in New Mexico.[6]

Ken Caldeira is an atmospheric scientist based at the Carnegie Institution at Stanford University, to which he moved in 2005 after ten years at the Lawrence Livermore National Laboratory. At Livermore he came under the influence of Lowell Wood, the legendary 'weaponeer' nicknamed 'Dr Evil'.[7] Caldeira too is fully aware of climate science and seems much more alarmed than Keith at the harm global warming will cause to humans. On the question of planetary management, though, his opinions seem to fluctuate. He says that thinking of geoengineering as a substitute for emissions reduction is 'crazy';[8] but he has been unable to pinpoint 'the one really bad thing that argues against geoengineering the climate'.[9] He is quoted as saying that sulphate aerosol spraying 'seems to be a dystopic world out of a science fiction story';[10] but also that 'I am not clear on what the "greenest" path is. Is it better to let the Greenland ice sheet collapse and let the polar bears drown their way to extinction, or to spray some sulphur particles in the stratosphere?'[11]

For some years Keith and Caldeira have been Bill Gates's principal source of expert knowledge on climate change.[12] From a series of briefings Gates has learned of the danger the world faces and what might be done about global warming. He was persuaded to commit several million dollars to finance research into geoengineering. The money is dispensed by Keith and Caldeira through the Fund for Innovative Climate and Energy Research.[13] Around half of the funds have been allocated to their own research but some have been used strategically to help finance a number of important meetings of the geoengineering community, including the Asilomar meeting (discussed later), the Royal Society processes, and workshops in Boston, Edinburgh and Heidelberg.[14]

Bill Gates is now the world's leading financial supporter of geoengineering research. He is an investor in Silver Lining, a company pursuing marine cloud brightening methods.[15] Blurring the boundary between disinterested research and financial reward that increasingly characterizes geoengineering, one of the more detailed research papers on marine cloud brightening names 25 authors of whom ten are affiliated with Silver Lining.[16] Gates is also an investor in Carbon Engineering Ltd, a start-up company formed by David Keith to develop technology to capture carbon dioxide from ambient air on an industrial scale. (Another investor in Keith's company is N. Murray Edwards, a Canadian oil billionaire with perhaps the largest financial stake in developing Alberta's tar sands, the worst source of fossil fuels, which is a little like a tobacco corporation donating to cancer research – and the cancer researchers accepting the money.)[17] Keith, who with others owns the patent for the carbon-sucking 'Planetary Cooler',[18] has said that if the right conditions can be found to construct his machines, 'we're printing money'.[19]

In addition to advising Gates and dispensing his research funds, Ken Caldeira is linked to Gates through a firm known as Intellectual Ventures, formed by former Microsoft employees and led by Nathan Myhrvold, once chief technology officer at Microsoft. Caldeira is listed as an 'inventor' at Intellectual Ventures.[20] Lowell Wood, once Myhrvold's academic mentor, retired from the Lawrence Livermore National Laboratory in 2007 to team up with Intellectual Ventures.[21] Gates is an investor. The company, whose motto is 'inventors have the power to change the world',[22] has developed the 'StratoShield', a hose suspended by blimps in the sky to deliver sulphate aerosols. The device is marketed as 'a practical, low-cost way to reverse catastrophic warming of the Arctic – or the entire planet'.[23] Intellectual Ventures has patented several geoengineering concepts, including an ocean pump for bringing cold seawater to the surface. That patent lists Caldeira, Myhrvold and Gates as inventors along with Lowell Wood and Roderick Hyde, co-authors with Edward Teller of the seminal paper on sulphate aerosol spraying.[24] Caldeira has said that if any funds come his way from these patents he will donate them to non-profit organizations.[25] To add to the incestuous impression, John Latham and Stephen Salter, the primary supporters of marine cloud brightening, are also named on the patent. The company defends its decision to privatize the intellectual property by arguing that a patent is the best way of communicating detailed technical ideas.[26] (Some would just go to the company's website to find them.) It also claims that monopolizing the knowledge will mean that their idea is 'better cared for'. More importantly, it claims, if Intellectual Ventures owns the patent to its stratospheric shield then Myhrvold and his associates can prevent the plan falling into the hands of those who cannot be trusted to use it responsibly.

Gates has been criticized for 'dissing' renewable energy and energy efficiency measures like household insulation as solutions to global warming.[27] He is less interested in 'old technologies', even those with proven capacity, and dismissed solar energy as 'cute'. He prefers to support 'innovative solutions', breakthroughs yet to be developed by whizz-kids, even though the experts agree the obstacle is not the absence of innovative alternatives to fossil fuels but the policies to ensure they are taken up. Gates sees climate change as a technical problem that requires some kind of 'killer app', magical thinking according to some.[28] Gates epitomizes an emerging force in the push for an engineered climate – behind the genuflecting to 'mitigation first', the lure of the technofix is irresistible.

For the time being, governments remain wary of geoengineering. Ken Caldeira coined the term 'solar radiation management' in the course of organizing a workshop with NASA in 2006 because bureaucrats were 'queasy' about using 'geoengineering'.[29]

> We were thinking what was the most boring and bureaucratic sounding name that we could make up that would let our workshop fly under the radar. I came up with 'Solar Radiation Management' as something that would obscure our meaning, while at the same time poke fun at the Washington-based bureaucratic mindset – I was laughing at the invention of 'Solar Radiation Management', thinking that this is the kind of boring obfuscatory language that DOE [Department of Energy] would use.

Subsequently, Caldeira decided that the word 'radiation' is too closely associated in the public mind with the dangers of nuclear power, so he began to translate the acronym SRM into the benign-sounding phrase 'sunlight reflection methods'. The term solar radiation management may have appeal to some anti-geoengineering activists, he has said, because 'it helps them sow confusion'. He does not accept that his joke on Washington bureaucrats backfired, or that switching to 'sunlight reflection methods' could be seen as a ham-fisted attempt at rebranding.

The emerging lobby

Richard Branson is another billionaire who hopes to save the world with a technofix. He sees himself as 'a committed crusader and ambassador of crucial and urgent social as well as environmental issues'.[30] Branson's 'Virgin Earth Challenge' has offered a $25 million prize to whoever can develop the best plan to extract carbon from the atmosphere. Of the 11 finalists in the competition four propose direct air capture methods and four are based on biochar.[31] Of more long-term significance, oil companies, anticipating a shift in the political landscape, are quietly backing research into geoengineering. Royal Dutch Shell is funding study of liming the seas through Cquestrate, an open-source, non-profit organization in Britain.[32] The chief scientist at the oil giant BP, Steven Koonin, was the convener of an expert meeting for the Novim Group, a non-profit scientific corporation, that in 2009 produced an influential report (considered later) on climate engineering as a response to climate emergencies.[33] The authors felt the need to declare that in playing a prominent role Koonin had no conflict of interest, arguing, implausibly, that it is not possible to say that promoting research into geoengineering has any bearing on policies to reduce carbon dioxide emissions and thus BP's bottom line. In 2009 Koonin was appointed Under-Secretary for Science at the United States Department of Energy.[34]

Despite ExxonMobil's long campaign to discredit climate science – in 2006 the Royal Society felt the need to write to the oil giant asking it to honour its promise to cease funding dozens of groups that have 'misrepresented the science of climate change by outright denial of the evidence'[35] – it too has now inserted itself into climate engineering. The corporation's point man on geoengineering is Haroon Kheshgi, who leads its Global Climate Change programme. A chemical engineer, Kheshgi was recruited to Exxon in 1986 from the Lawrence Livermore National Laboratory. He is equivocal about whether human-induced climate change is anything to worry about.[36] In 1995, ensconced at Exxon, he was the first to propose liming the oceans as a means of reducing acidification due to escalating atmospheric carbon.[37] Through Kheshgi, Exxon has begun to influence 'independent' reports into geoengineering, such as the 2007 NASA report on solar radiation management organized by Caldeira. The oil company also funded a report (considered in the next chapter) concluding that sulphate aerosol spraying would be a much cheaper response to global warming than phasing out fossil fuels. Its CEO, Rex Tillerson, has described climate change as an 'engineering problem' with 'engineering solutions'.[38]

A range of companies have identified business opportunities in geoengineering. Some believe they can profit from carbon credits because polluters with emission caps will pay them to take carbon dioxide out of the atmosphere, even though there is no chance of this kind of activity being recognized in emissions trading systems in the foreseeable future.[39] One of the distinctive features of solar radiation management methods of geoengineering is that no one can make money out of trading emission reduction credits, although some, like those at Intellectual Ventures, believe the world may need to pay a great deal for access to patented technology that can prevent climate catastrophe. In 2007 Planktos, a company set up by an entrepreneur with a colourful background and financed by a Canadian real estate developer, announced plans to fertilize the oceans near the Galapagos Islands.[40] With an eye to the growing market for carbon offsets, Planktos's environmental claims were plausible enough for gullible investors to raise the value of the company to $90 million. One broker flogging Planktos Corp. stock urged his customers to 'Get in on the groundfloor NOW before . . . $8,000 transforms into a whopping $192,000!' As the ship set sail the responsible authorities got wind of it. The voyage was stopped and the venture collapsed, leaving a cloud of mistrust hanging over all research into iron fertilization. One of the benefits of the Planktos debacle was that it alerted regulators to the dangers of rogue geoengineers and the absence of a regulatory framework for all research.[41] Not long after, both the London Convention, which regulates ocean dumping, and the Convention on Biological Diversity passed resolutions banning iron fertilization experiments except under restricted conditions.

Regulation moves more slowly than commerce and in recent years there has been a flurry of patents taken out over methods to engineer the climate. Some of them are so broad that, if enforceable, they would place fertilization of the oceans in the hands of one man.[42] Another man holds a patent with the description: 'Use of artificial satellites in earth orbits adaptively to modify the effect that solar radiation would otherwise have on Earth's weather.' Most people would assume that if anyone is to own a technology to control the amount of sunlight reaching the planet it should be government, rather than Franklin Y. K. Chen of One Meadow Glen Rd, Northport, New York. The US Patent and Trademark Office disagrees. So does Ken Caldeira, who cannot see that patenting a solar shield is any different from patenting a new drug.[43] In 2010 Shobita Parthasarathy and co-authors noted a sharp increase in geoengineering patents in recent years and warned that, as in the case of biotechnology, the patents owned by private companies and individuals are on track to become the de facto form of governance of geoengineering.[44] We are approaching a situation in which international efforts to protect humanity from climate catastrophe could depend on whether or not one company wants to sell its intellectual property.

It is clear, even at this early stage, that burgeoning commercial engagement in geoengineering is creating a constituency with an interest in more research and, eventually, deployment. Such a lobby is naturally predisposed to argue that pursuing mitigation is 'unrealistic' or 'politically impossible' and that therefore climate engineering is the sensible alternative, if only so that we can 'be prepared'. Already the chorus of demands for public funding of research is loud; every inquiry and report calls for large amounts of it. It is fair to expect that if we reach the stage of deployment, any move to terminate it (due, for example, to evidence of unexpected environmental damage) would be fought by the new industry with complaints of asset devaluation and job losses. Today it may seem absurd that factors like these should play a role in deciding the fate of the entire planet, but the history of environmental policy-making shows that these kinds of decisions are never based solely on public safety. As President Eisenhower warned in 1961, there is a 'danger that public policy could itself become the captive of a scientific-technological elite'.[45]

It is easy to see this happening in the emerging politics of the Arctic. The melting of summer sea-ice has triggered a 'new gold rush' to secure control over the newly accessible oil and mineral resources. China, Brazil and India are lobbying to join the Arctic Council where decisions are made about who gets access to what.[46] According to an energy industry insider:

Climate change is opening up one of the last frontiers for hydro-carbons on our planet. The Arctic could hold around 25% of undiscovered oil and gas reserves and the fact that the ice is retreating for whatever reason means that the region could be set for rapid change and development as exploration, produc-tion and infrastructure will have an inevitable, irreversible impact.[47]

In a provocative move in 2007 a Russian submarine planted its national flag on the seabed under the North Pole. Other Arctic states – notably Canada, USA and Norway – have invested in new Arctic military capability.[48] Norway has shifted its army headquar-ters from Oslo to the northern town of Bardufoss, with a new fleet of jets to be based there.[49] In 2008 Russian President Dmitry Medvedev declared: 'Our first and main task is to turn the Arctic into Russia's resource base of the 21st century'.[50]

While shrinking ice reveals a new El Dorado, and remilitariza-tion of the Arctic proceeds apace, scientists are proposing schemes to target the Arctic with stratospheric aerosol spraying in order to rebuild ice. The call by Mike MacCracken of the Climate Institute (a Washington environmental NGO) for 'aggressive research' into reversing Arctic warming and returning ice by aerosol spraying has attracted support.[51] Intellectual Ventures is pushing its proprietary solar shield specifically to counter Arctic warming. As the ice retreats over the next two to three decades we may be confronted with a climate emergency in which calls for urgent rebuilding of Arctic ice are opposed by an energy industry dependent on easy access to the Arctic sea-floor and nations such as Russia whose energy security rests on thin ice.

In the same way, suppliers of sulphur would acquire a commer-cial interest in climate control via sulphate aerosol spraying because it would require a large and continuous supply of the chemical. Most sulphur used today is a by-product of oil and gas production, although a share is scrubbed out of the flue gases of coal-fired power plants. (It would be the final irony if we were to extract sulphur before it pollutes the lower atmosphere only to pump it into the upper atmosphere to prevent climate change.) The sulphur industry already has its own lobby group in Washington.[52] Dominated by oil companies, it actively promotes greater use of sulphur. Once major corporations have a stake they become a political force with an interest in growth. It is perhaps not too far a stretch to suggest that, if a programme of sulphate spraying were begun and accounted for 5 or 10 per cent of world sulphur demand, lobbying to protect the shareholder value of sulphur producers could influence decisions about the planet's climate.[53]

Who knows about geoengineering?

What does the public think about all of these developments? There is almost no information on public attitudes to geoengineering for the simple reason that almost no one has yet heard of it. A study in 2009 found that only 3 per cent of Americans could correctly describe geoengineering; 74 per cent had not heard of it at all and the rest were 'wildly misinformed'.[54] In March 2011, a large online survey in Japan found that only 7 per cent of the adult population had heard of 'geoengineering' and 11 per cent of 'climate engineering'.[55] The figures are unlikely to be much different elsewhere.[56] Nevertheless, social scientists are beginning to investigate public attitudes to geoengineering. Given that only few among the public have heard of it, an approach based on focus groups and deliberative workshops is necessary. At these gatherings the main technologies and issues are explained and reactions are then explored.

Conscious of the power of 'framing', the organizers of deliberative workshops in Britain, Karen Parkhill and Nick Pidgeon, set out to inform participants about geoengineering in a neutral way. They were aware that perceptions of 'naturalness' could sway assessments of various technologies.[57] Antipathy to technological interference in nature's complex systems and preference for the 'natural' were among the findings of a public dialogue on geoengineering carried out by the Natural Environment Research Council in Britain.[58] The naturalness test works against stratospheric aerosol spraying and in favour of some forms of carbon dioxide removal, such as biochar and reforestation. As they learned about the technologies, the workshop participants tended to see stratospheric aerosol spraying as deepening the alienation of humans from nature. And they saw it as a 'stop-gap' rather than a solution to the problem. Against this, the survey in Japan referred to above found unexpected levels of support for geoengineering, once it had been explained. Half supported use of geoengineering to combat global warming, although nine out of ten agreed that 'Earth's temperature is too complex for a single technology'.[59]

As public recognition of geoengineering is embryonic, early media framing by advocates and critics will set public attitudes. Critics are likely to emphasize the complexity, unintended effects and artificiality of solar radiation management, similar to the campaign against genetically modified organisms. The term 'Frankenclimate' has already entered the lexicon.[60] Proponents are increasingly presenting climate engineering as an emergency response, perhaps preparing to sideline public opinion. 'You may not like it, but we have no choice.' The organizers of the deliberative workshops have attempted to avoid framing the need to engineer the climate as an emergency response because they understand how it disempowers the public and undercuts dissenting positions.[61]

Nevertheless, every presentation comes with a frame. Holly Buck observes that when well-intentioned scientists tell focus groups that we 'need to trade-off pros and cons of different technologies' participants are drawn into a rationalistic, economic way of thinking.[62]

In an early sign of public and official nervousness about solar radiation management, in September 2011 British scientists announced the postponement of a trial of a geoengineering technology, part of a larger project known as SPICE, Stratospheric Particle Injection for Climate Engineering. The project is the initiative of four British universities, with technical support from Marshall Aerospace, a company specializing in the modification and maintenance of aircraft. The plan was to suspend a hose 1 kilometre long using a helium-filled blimp to investigate the suitability of certain particles, to test the effectiveness of delivery systems (hose, nozzles, balloon and tethers) and to monitor potential impacts on weather and atmospheric chemistry.[63] The performance of balloons in high winds is an early research question. Although involving a hose only 1 kilometre in length and spraying only water, the proposed trial attracted strong criticism from a range of civil society organizations coordinated by the Canada-based ETC Group.

In May 2012 the trial was cancelled after it became known that one of those who recommended it for funding owned a patent in the hose suspension system. The difficulties encountered by the SPICE project have heightened wariness among geoengineering researchers, not least because that project appeared to have strong governance arrangements, even if its public communication strategy failed. The response is unlikely to be a scaling back of geoengineering research and testing; rather, funding and approval for geoengineering research is likely to be secured under a different name, since most geoengineering research can be presented less controversially.

Cultures of denial

When people hear about geoengineering, their attitudes are formed in part by the way it is presented and in part by the values and worldview they bring to it. Already we have seen that some become excited at the idea of a technofix, while others are suspicious of any plan to control nature by artificial means. Crucially, attitudes to geoengineering will be heavily influenced by beliefs about climate change. Geoengineering research has blossomed because those involved became convinced that the overwhelming evidence that the world is warming dangerously is not enough to overcome the political, psychological and cultural barriers to cutting emissions. Perhaps the foremost reason for the reluctance to reduce global greenhouse gas emissions (and so the turn to geoengineering) has been the conservative backlash against climate science and climate policy in the United Sates. Some understanding of climate denial is essential, therefore, not only to grasp why the world has failed to respond to the scientific warnings with alacrity but also to explain why, as we will see, there is a growing interest in climate engineering even among those who do not accept that global warming is occurring. If forms of denial structure the interpretation of a problem, they will also frame thinking about the solutions to it.

In the United States the repudiation of science – which began as a self-interested industry campaign in the 1990s and morphed into a much more powerful political and cultural movement in the 2000s – has reached the highest political levels. In April 2011 the following proposition was put to the US House of Representatives: 'Congress accepts the scientific findings . . . that climate change is occurring, is caused largely by human activities, and poses significant risks for public health and welfare.' The House, dominated by Tea Party Republicans, voted by 240 to 184 to reject the basic propositions of climate science. It was as if American legislators had a mandate to vote down the laws of physics. A few months earlier the US National Academy of Sciences had published a major review of the scientific literature before concluding: 'A strong, credible body of scientific evidence shows that climate change is occurring, is caused largely by human activities, and poses significant risks for a broad range of human and natural systems.'[64]

In the United States, then, politics has defeated science. The triumph of ideology followed a long and successful campaign to open up a gap between the views of liberals and conservatives on global warming. In 1997 there was virtually no difference, with around half of Democrats and Republicans saying warming had begun. In 2008, reflecting the accumulation and dissemination of scientific evidence, the proportion of Democratic voters taking this view had risen from 52 per cent to 76 per cent.[65] But the proportion of Republican voters fell from 48 per cent to 42 per cent – a 4 per cent gap had become a 34 per cent gap. What had happened?

The opening of the gulf was due to the fact that Republican Party activists, in collaboration with fossil fuel interests and conservative think tanks, had successfully characterized those accepting global warming science as 'liberals', a term of abuse for American conservatives.[66] In other words, they had activated the human predisposition to consolidate one's identity by cementing one's connection with cultural groups.[67] In the 1990s views on global warming were influenced mostly by attentiveness to the science; now one can make a good guess at an American's opinion on global warming by identifying their views on abortion, same-sex marriage and gun control.

Liberals may be as predisposed as conservatives to sift evidence through ideological filters (although there is some truth in the conservative accusation that liberals are 'self-doubting', which to a liberal generally means 'more thoughtful and self-critical'); but in the case of global warming it happens that the evidence overwhelmingly endorses the liberal beliefs that unrestrained capitalism is jeopardizing future well-being, that comprehensive government intervention is needed, and that the environment movement was right to raise the alarm about global warming. For many conservatives, accepting these reliefs is intolerable; it is easier emotionally and more convenient politically to reject climate science.

Naomi Klein has argued that while the deniers are completely wrong on the science, they understand much better than liberals the political implications of accepting the science – 'the kind of deep changes required not just to our energy consumption but to the underlying logic of our economic system'.[68] The irony is that this was much less true in the 1990s; if the transformation of energy infrastructure had begun in earnest in the 1990s, then, according to all of the economic assessments, the costs would have been very low, the kind of structural change to one sector that has been managed before, and we would now be well on the way to low-emissions economies. The delay caused by the early campaign of obstruction by the fossil fuel corporations and their acolytes ramped up the economic costs of climate protection and turned climate change into a deep and unbridgeable cultural divide.

The aggressive adoption of climate denial by neoconservatism was symbolized by the parting gesture of George W. Bush at his last G8 summit in 2008. As he left the room he turned to the assembled leaders to say: 'Goodbye from the world's biggest polluter.'[69] It was a defiant 'joke' reflecting the way US neoconservatives define themselves by their repudiation of the 'other', in this case, the internationalist, environmentally concerned, self-doubting enemies of 'the American way of life'. Conceding ground on global warming would have meant bridging two implacably opposed worldviews. Bush's words, and the fist pump that accompanied them, were read by those present as a two-fingered salute to everything the Texan opposed. Yet today's Tea Party Republicans look back on George W. Bush as too soft and accommodating.

With the patronage of *Fox News*, the Tea Party movement took off in early 2009. So powerful has its influence been on conservative politics that in early 2012 all six Republican presidential candidates had repudiated their previous support for carbon abatement measures and rejected the science of climate change. As governor of Massachusetts, Mitt Romney was one of the most vigorous advocates of emission reductions. When he ran successfully for the nomination, a campaign funded substantially by fossil fuel companies, he renounced climate science.[70] In 2008 even Sarah Palin had expressed concern about global warming and called for a cap-and-trade system. By the end of 2011, Jon Huntsman, the lone Republican presidential candidate who had defended climate science and warned that the Republicans were becoming the 'anti-science' party, had buckled under the pressure, questioning the validity of the science and suggesting scientists might be fabricating evidence.[71] As one despairing, Republican-voting meteorologist put it, climate science has become 'a bizarre litmus test for conservatism'.[72]

It is now well established statistically that in the United States denial of climate science is much more common among conservative white males than other demographic groups.[73] It is also true that those white males who feel most confident and knowledgeable about climate science are more likely to deny the existence of anthropogenic warming. Thus, 49 per cent of conservative white males who believe they understand the issue well say global warming will never happen; among less 'well-informed' conservative white males only 19 per cent take the same view. While 57 per cent of 'well-informed' conservative white males say they are not at all worried about global warming, only 29 per cent of their less well-informed allies and 14 per cent of all other adults feel so relaxed.[74]

The authors of the study that uncovered these facts, Aaron McCright and Riley Dunlap, argue that they can be explained by some well-known sociological and psychological phenomena. First, compared to other adults white males have been shown to be less averse to a wide range of risks. This may be due to the fact that traditionally, as the dominant social group, they are actually less subject to social and environmental threats and more able to control their environment. Alternatively, I'd suggest, their sense of identity may have been destabilized by the deep changes flowing from the social revolutions of the 1960s and 1970s, which saw a 'cultural backlash' in films and television programmes. In addition, for many years the most public voices of climate science denial in the media (*Fox News* commentators, Rush Limbaugh, spokesmen from right-wing think tanks and so on) have been high-status conservative white males with whom other conservative white males identify.

If, as McCright and Dunlap claim, 'conservative white males are likely to favor protection of the current industrial capitalist order which has historically served them well',[75] the key word here is 'order', the felt need to maintain the stability of the social system. In the past, threats to the established order have come from various political actors – socialists, feminists, environmentalists, Islamists. Now the threat is from a disturbed natural world. It is pointless to rail against the weather, so those fearful of destabilization have displaced the problem onto those who announce it – scientists and political leaders who say we must change our ways.

We should not push the white male effect too far. While the tendency to adopt climate denial is especially strong among conservative white males, other factors are at work. Denial in various forms is widespread in the rest of the population. And the survey results indicate that some 30 per cent of those identified as confident conservative white males *do* believe there is a scientific consensus on global warming, and 43 per cent of them *do* worry about a changing climate.

Nevertheless, these results are consistent with the more general argument that conservatives tend to take a more hierarchical view of society, as a natural order in which some groups are dominant and some subservient. They are more likely to accept that exercising control and authority, over both society and the environment, are natural and desirable because the alternative means disorder and indecision. This kind of analysis goes a long way towards explaining an apparent paradox – organizations such as the Heartland Institute and the American Enterprise Institute that actively reject climate science also support geoengineering.[76] The co-director of the American Enterprise Institute's Geoengineering Project writes that geoengineering 'is the most revolutionary and potentially valuable new idea in climate policy today .... [and] challenges us to take the climate seriously'.[77] Like the patient who will accept the doctor's diagnosis only if the illness is treatable, a solution to global warming that does not destabilize a person's worldview – but in fact vindicates it – makes recognizing the problem palatable.

The white male effect has far-reaching implications for the imminent debate over geoengineering. Consider what we know. Conservative white males have more invested in defending the prevailing social order. Compared to 'ruinous' carbon abatement policies, as a technofix geoengineering promises to protect the structure of economic and political power, the balance sheets of the fossil fuel corporations, unfettered markets, and the 'rights' of consumers. As the identity of conservative white males tends to be more strongly bound to the prevailing social structure, geoengineering is the kind of solution to climate change that is less threatening to their values and sense of self. Whereas mitigation policies like carbon taxes and emissions trading activate resistance from 'system-justifying attitudes', those same attitudes are likely to trigger support for geoengineering solutions because they are consistent with ideas of control over the environment and the personal liberties associated with free market capitalism. Just as the need to defend a cultural worldview makes conservative white males prone to repudiate climate science, so that worldview will make them prone to support geoengineering solutions. Instead of climate change jeopardizing the system with which they identify, geoengineering would represent the triumph of 'man over nature'. Technological intervention, in which they have an unusual degree of confidence, reaffirms human technological mastery.

Work by Dan Kahan and others has shown how cultural values are cognitively prior to facts so that, if they are to be accepted, facts must accommodate those values.[78] Evidence that contradicts understandings of how society should be organized is liable to be dismissed: 'taking a position at odds with the dominant view in his or her cultural group is likely to compromise that individual's relationship with others on whom that individual depends for emotional and material support.'[79] Previous work has shown that individuals with individualistic and hierarchical values are prone to dismiss the risk of global warming because accepting the danger would be an indictment of social elites they respect, and require limits on the free market. Those with egalitarian and communitarian values are more inclined to accept that there is a problem because the answer to it does not provoke resistance. In an empirical study Kahan and his co-authors hypothesize that the option of responding to climate change with geoengineering – which symbolizes the ability of humans to respond with technological creativity to a problem that business and the market seems to have created – would make those with individualistic-hierarchical values less dismissive of the facts of climate science. Although the results should be generalized with caution, the study found that the geoengineering option does indeed make them somewhat less likely to dismiss the science. The effect is much less pronounced in Britain than in the United States, which is unsurprising given that polarization has not become extreme in the former.

All of this adds some empirical substance to a conclusion that has been apparent for some time. What is astonishing about the paper by Kahan and his co-authors is the conclusion they draw from it: because geoengineering promises to reduce cultural polarization by affirming the values of hierarchical individuals it offers a way out of the political impasse over climate change. 'In order to overcome cultural resistance to sound scientific evidence ... people of diverse values must all be shown *solutions* that they find culturally congenial.'[80] In other words, if planning to take control of the Earth's climate system forever – using highly speculative technologies fraught with political and scientific uncertainties and risks – is needed to appease those whose prejudices prevent them from accepting scientific facts, then that is what we must do. Here the researchers, in their good liberal pursuit of compromise and 'fixing the communications failure', seem to slide into a post-modern view that scientific facts are always conditioned by culture and must be subordinated to it no matter the cost.

We know we are in trouble when liberals who say they accept the science begin advocating geoengineering as a means of appeasing conservatives who reject the science. Once the authority figures trusted by conservative white males begin to promote the benefits and depreciate the risks of geoengineering we can expect them to swing firmly behind it and to set out to acquire information about the technologies with a view to confirming their biases in favour of it. We can anticipate that a pro-geoengineering coalition will form between those conservative white males who reject climate science and those who accept it. Together they are more likely to promote grand, system-altering interventions like sulphate aerosol spraying than more benign, 'soft' geoengineering technologies like reforestation, biochar and painting roofs white. And they will take the view that if we are going to geoengineer the planet then it must be the United States, perhaps in coalition with trusted allies, that controls the process and not the United Nations, which they tend to hold in contempt.

Denial around the world

In the United States denying climate science is not simply a form of oppositionism; it is a powerful energizing force for conservatism. Rejecting science means rejecting all measures based on it, even if they are otherwise desirable. Thus conservatives linked to the Tea Party have mounted vigorous campaigns against public transport, solar energy, bicycle lanes and smart meters, claiming that local conservation initiatives are in fact part of a conspiracy against the 'American way of life' promoted by the United Nations' Agenda 21, a well-meaning and non-binding set of principles agreed at the Rio Earth Summit in 1992.[81] It would be natural to dismiss these moves as paranoid farce – one Tea Party activist was applauded when she claimed that the 'real job of smart meters is to spy on you and control you' – except that officials in many areas have been spooked by them. Political actors who would normally have no truck with this kind of foolishness are playing to the fears. In early 2012 the Republican Party convention passed a resolution condemning 'the destructive and insidious nature' of Agenda 21. In June 2012 the state Senate in North Carolina voted overwhelmingly for a law that bans all government agencies from making plans or developing regulations that take account of future sea-level rise. In case the reader thinks this childlike refusal to see what we don't like is confined to backward legislators in 'fly-over states' in America, in Australia Rupert Murdoch's flagship newspaper, the *Australian*, has for years been ridiculing scientific warnings of sea-level rise, countering the 'alarming predictions' of research published in professional journals with the common-sense observations of long-term beachgoers.[82]

These campaigns are bewildering until we remember that resisting environmental regulations, including energy efficiency measures, has become a sign of red-blooded faith in the prevailing system, the particular conservative construction of the American way of life. The ideological framing of environmentally benign technology has a long history in the United States. When Sherwood Rowland, the American chemist who would share the 1995 Nobel Prize with Paul Crutzen, advocated a ban on ozone-destroying consumer products, the aerosol spray-can industry suggested he was a KGB agent bent on destroying capitalism.[83] While renewable energy industries in the United States face constant political attacks, in China investment in green technologies is surging. It would be a paradox of history if it turned out that democracy in America had become so dysfunctional that it could be held hostage by an anti-environmental minority while a totalitarian government in China took decisive action on the threat of global warming, and in the process assumed the mantle of world leadership in which an emergent 'Chinese way of life' proved superior to its American counterpart.

Even so, it seems clear that the United States is at a stage in its history where it is having difficulty making good decisions in its own long-term interests, let alone those of the the rest of the world. The era in which judgements must be made about geoengineering has begun; within two or three decades a momentous choice will need to be made about deployment of Earth-changing technologies. Although the United States is not short of intelligent, thoughtful and deeply concerned people, from today's vantage point it is hard to see it regaining enough political composure to be able to reflect carefully on the implications of its choices.

Climate denial has spread, often through coordinated effort, from the United States to other Anglophone countries that have similarities in their political cultures, especially to Australia, Britain and English-speaking Canada. Influenced by denialism, the conservative government in Canada pulled out of the Kyoto Protocol and vigorously promotes development of tar sands. In Britain, despite the prominence given to Lords Monckton and Lawson by the right-wing press and a bevy of virulent columnists, the Conservative Party presents itself as 'greener' than the Labour Party, so denial is left to the far-right British National Party, whose impact is negligible. In Australia, where conservatives have waged a culture war and denialism has put down roots, the conservative party (known as the Liberal Party) is now dominated by those who reject climate science but pretend otherwise for electoral reasons. In continental Europe the absence of a long-running and rancorous culture war explains the relative weakness of climate denial there. In France, a book by scientist and former science minister Claude Allègre denying climate science gained a short-lived notoriety in 2010 but denialism has not taken root. Although *L'Imposture climatique* (*The Climate Fraud*) sold over 100,000 copies, its 'factual mistakes, distortions of data, and plain lies' left Allègre (who had previously claimed asbestos was harmless) with little credibility.[84] In Italy and some former Eastern bloc countries, where anti-communism still influences right-wing politics, denial is more potent, but its credibility was damaged by association with Silvio Berlusconi. In Germany, climate denial is invisible; perhaps that nation is all too conscious of the perils of twisting science to suit ideological ends, perils illustrated by a surprising historical parallel.[85]

It is hard to imagine a scientific breakthrough more abstract and less politically contentious than Einstein's general theory of relativity. Yet in Weimar Germany in the 1920s it attracted fierce controversy, with conservatives and ultra-nationalists reading it as a vindication of their opponents – liberals, socialists, pacifists and Jews. They could not separate Einstein's political views – he was an internationalist and pacifist – from his scientific breakthroughs, and his extraordinary fame made him a prime target in a period of political turmoil. There was a turning point in 1920. A year earlier a British scientific expedition had used observations of an eclipse to provide empirical confirmation of Einstein's prediction that light could be bent by the gravitational pull of the Sun. Little known to the general public beforehand, Einstein was instantly elevated to the status of the genius who outshone Galileo and Newton.[86] But conservative newspapers provided an outlet for anti-relativity activists and scientists with an axe to grind, stoking nationalist and anti-Semitic sentiment among those predisposed to it. In a similar way today, conservative news outlets promote the views of climate deniers and publish stories designed to discredit climate scientists, all with a view to defending an established order seen to be threatened by evidence of a warming globe. As in the Weimer Republic, the effect has been to fuel suspicion of liberals and 'elites' by inviting the public to view science through political lenses.

At the height of the storm in 1920, a bemused Einstein wrote to a friend: 'This world is a strange madhouse. Currently, every coachman and every waiter is debating whether relativity theory is correct. Belief in this matter depends on political party affiliation.'[87] The controversy was not confined to Germany. In France a citizen's attitude to the new theory could be guessed from the stance he or she took on the Dreyfus affair, the scandal surrounding the Jewish army officer falsely convicted of spying in 1894, whose fate divided French society. Anti-Dreyfusards were inclined to reject relativity on political grounds.[88] In Britain, suspicions were less politically grounded but relativity's subversion of Newton was a sensitive issue, leading Einstein to write an encomium for the great English scientist prior to a lecture tour.

Like Einstein's opponents, who denied relativity because of its perceived association with progressive politics, conservative climate deniers follow the maxim that 'my enemy's friend is my enemy', so scientists whose research strengthens the claims of environmentalism must be opposed. Conservative climate deniers often link their repudiation of climate science to fears that cultural values are under attack from 'liberals' and progressives. In Weimar Germany the threat to the cultural order apparently posed by relativity saw Einstein accused of 'scientific dadaism', after the anarchistic cultural and artistic movement then at its peak. The epithet is revealing because it reflected anxiety that Einstein's theory would overthrow the established Newtonian understanding of the world, a destabilization of the physical world that mirrored the subversion of the social order then underway. Relativity's apparent repudiation of absolutes was interpreted by some as yet another sign of moral and intellectual decay. There could not have been a worse time for Einstein's theory to have received such emphatic empirical validation than in the chaotic years after the First World War.

Although not to be overstated, the turmoil of Weimar Germany has some similarities with the political ferment that characterizes the United States today – deep-rooted resentments, the sense of a nation in decline, the fragility of liberal forces, and the rise of an angry populist right. Environmental policy and science have become battlegrounds in a deep ideological divide that emerged as a backlash against the gains of the social movements of the 1960s and 1970s.[89] As we saw, marrying science to politics was a calculated strategy of conservative activists in the 1990s,[90] opening up a gulf between Republican and Democratic voters over their attitudes to climate science. Both anti-relativists and climate deniers justifiably feared that science would enhance the standing of their opponents and they responded by tarnishing science with politics.

Einstein's work was often accused of being un-German, and National Socialist ideology would soon be drawing a distinction between Jewish and Aryan mathematics.[91] Although anti-Semitism plays no part in climate denial, 'Jewish mathematics' served the same political function that the charge of 'left-wing science' does in the climate debate today. In the United States, the notion of left-wing science dates to the rise in the 1960s of what has been called 'environmental-social impact' science which, at least implicitly, questioned the unalloyed benefits of 'technological-production' science.[92] Thus in 1975 Jacob Needleman could write: 'Once the hope of mankind, modern science has now become the object of such mistrust and disappointment that it will probably never again speak with its old authority.'[93] The apparent paradox of denialist think tanks supporting geoengineering solutions to the global warming problem that does not exist can be understood as a reassertion of technological-production science over environmental impact science. Thus the Exxon-funded Heartland Institute – the leading denialist organization that has hosted a series of conferences at which climate science is denounced as a hoax and a communist conspiracy – has enthusiastically endorsed geoengineering as the answer to the problem that does not exist.[94] The association between 'left-wing' opinion and climate science has now been made so strongly that politically conservative scientists who accept the evidence for climate change typically withdraw from public debate, as do those conservative politicians who remain faithful to science.

The motives of Einstein's opponents were various but differences were overlooked in pursuit of the common foe, just as today among the enemies of climate science we find grouped together activists in free market think tanks, politicians pandering to popular fears, conservative media outlets like the *Sunday Times* and *Fox News*, a handful of disgruntled scientists, right-wing philanthropists including the Scaifes and Kochs, and sundry opportunists such as Christopher Monckton and Bjorn Lomborg.[95]

Einstein's theories were not taught in German schools in the 1920s; had they been it is likely anti-relativists would have mounted a campaign to demand teachers stick to 'pure science'. In May 2010 the *Denver Post* reported the formation of a new group called Balanced Education for Everyone, whose purpose is to stop the teaching of climate science in American schools.[96] Dismissing it as 'junk science', the group claims teaching children about global warming is wrong and frightens them. In Oklahoma a bill titled the Scientific Education and Academic Freedom Act was introduced to the legislature that would require teachers to present 'both sides' of the debates over climate science and evolution.[97] In 2011 a school board in California became nervous about climate science appearing in its curriculum and adopted a policy of requiring teachers to teach it in a 'balanced' way.[98] The school board had fallen for the fiction, deliberately created by denier organizations, that there is substantial disagreement among climate scientists about global warming and its causes. As in the case of evolution, science teachers are now being accused by scientifically illiterate parents and administrators of forcing their 'beliefs' on pupils. A Wisconsin teacher said he faced a 'lynch mob' mentality after the local Tea Party invited teachers to a debate with prominent climate deniers in front of students from 200 local high schools. The organizers described climate science as a 'monstrous hoax'.[99] The Heartland Institute mailed a pamphlet attacking climate science to 14,000 school boards. In Australia, prominent denier Ian Plimer published a book, promoted by a think tank with links to the mining industry and launched by former prime minister John Howard, claiming that when children are taught about climate change they are being fed propaganda. The book lists 101 questions with which children can challenge their science teachers.[100] In the past, Plimer's claims have been comprehensively, almost embarrassingly, debunked.[101] Any student who relied on them would fail their science courses. Plimer has been embraced by mining billionaire Gina Rinehart. In July 2012 in Queensland, an Australian state with a reputation for backwardness and still at times referred to as 'the deep north', the conference of the ruling conservative party (known as the Liberal National Party) passed overwhelmingly a resolution to ban teaching of climate science in government schools.[102] Declaring climate science 'environmental propaganda', the mover of the resolution described climate scientists as 'false prophets' who are poisoning young minds. A 'moderate' who spoke against the motion argued that children should be given all sides of the debate!

Despite its extraordinary political success, the denial movement is a long way from winning over the majority to its views,
even in the United States. While the rise of anti-science is disconcerting, it is more difficult to understand why the broader public
has not been demanding far-reaching measures consistent with the
scientific warnings. Although most members of the public –
including in the United States – accept the scientific consensus, by
sowing doubt the denial movement has provided reasons to accept
it with less conviction. Psychological research has identified the
various 'coping strategies' we might use to defend against or
manage the unpleasant emotions associated with the dangers of a
warming globe – fear, anxiety, anger, depression, guilt and helplessness.[103] Instead of repudiating the science outright, the coping
strategies admit some of the facts and allow some of the associated
emotions, but only in distorted form, such as turning fear into
apathy, or guilt into blame-shifting.

One strategy is a kind of jaded cynicism, caustically illustrated
in Ian McEwan's novel *Solar*. The leading character, Michael
Beard, is a brilliant but bored physicist approaching retirement.
Convinced of the threat posed by global warming he starts a
company to commercialize a radically new way to generate clean
energy from the Sun. A day before the grand public demonstration
of the new technology, Michael's business partner confides that
he is having serious doubts about their commercial prospects
because he has seen scientists on television saying the planet is
cooling. 'If the place isn't hotting up, we're fucked,' he says. Michael
patiently explains the errors of the deniers and reassures him
that the science is robust, adding: 'It's a catastrophe. Relax!'[104]
This could also become the catchphrase of those venture capitalists now investing in geoengineering technologies.

Another strategy of disavowal is to place climate change in its own mental compartment, beyond one's concerns, an approach that works most effectively in the higher reaches of intellectual life. So, for instance, in his recent book Nobel laureate economist Michael Spence is buoyant about continued economic growth in China, India and many poor countries. He foresees only two threats to a rosy future – a return to protectionism and the risks of fiscal deficits.[105] He mentions global warming only to wave it away with the observation that he is 'optimistic' about it. That someone of Spence's stature can remain impervious to the warning of every science academy in the world – that climate change will transform the conditions of economic life in the twenty-first century – is a mystery left for future historians to ponder.[106]

Other widely used coping strategies aim to ignore or belittle the problem.[107] Some people cultivate indifference to global warming and its implications. Apathy is typically understood as meaning the absence of feeling, but it can often reflect a suppression of feeling that serves a useful psychological function.[108] Others restrict their exposure to upsetting information or view it through a cloud of doubt. Who at times has not thought: 'If I don't care, I won't feel bad'? Many engage in what might be called 'casual denial'. Less vociferous than outright denial of the science, casual denial relies on inner narratives, such as 'Environmentalists always exaggerate' and 'I'll worry about it when the scientists make up their minds'. The desire to disbelieve is activated by conservative news outlets each time they give undue prominence to stories that create the impression that climate scientists cannot agree or that the science is politically tainted. Others blunt the emotional impact of the scientific warnings by emphasizing the time lapse before the consequences of warming are felt, or by believing that poor people in distant lands will be the victims.

Those who have adopted strategies that make climate change seem to be of minor consequence or irrelevant are unlikely to pay much attention to geoengineering. If they are not worried about climate change they are unlikely to be worried about proposed solutions to it, unless it affects them directly, which explains why so much resentment can be whipped up against a modest carbon tax. On the other hand, it is possible that the reality of climate control, with its far-reaching (not to say, frightening) implications, may bring home the reality of global warming. One geoengineering researcher has ventured the opinion that developing technologies such as sulphate aerosol spraying is like holding up to the public the instruments of torture.

Blame-shifting attempts to respond to feelings of guilt and anger by disavowing responsibility for the problem or the solution. It is a tactic in play when Australian governments stress that Australia contributes only 1.5 per cent of global emissions, even though Australians have the highest emissions per person in the industrialized world. Kari Norgaard writes of the residents of a town in Norway who shift blame for global warming onto 'Amerika' when they are reminded that Norway has grown rich by becoming one of the world's biggest oil exporters.[109] Blame is shifted by selective emphasis on certain facts, in order to allow one to say with a shrug, 'Well, what can I do?' Geoengineering inverts the situation because it is perfectly feasible, or soon will be, for a mid-sized nation to undertake it unilaterally. China or 'Amerika' may be those most responsible for the problem, but it will be in the power of citizens in many nations to 'solve' it. By providing an answer to 'What can I do?', the development of geoengineering technologies will deny us the resigned shrug.

Perhaps the most widespread and powerful defence against the unpleasant emotions triggered by climate science is wishful thinking. Cultivating 'benign fictions' can be comforting in an often unfriendly world, yet such fictions become dangerous delusions when they are clung to despite overwhelming evidence.[110] The climate debate is rife with wishful thinking, deploying narratives such as 'Technology will save us' (carbon capture and storage, nuclear power and biochar), 'We've solved these problems before and we will do it again', or simply 'Something will come along'. Geoengineering appeals to the wishful thinker because it can be presented as a magic solution to climate change; it is exactly the kind of solution in which we want to invest our hopes.

Not long ago in Cambridge I gave a talk on the various subterfuges we use to evade the full meaning of the scientific warnings.[111] In question time and conversation afterwards it became apparent that most seemed to have heard nothing I had said, for each of the ruses I had identified was unselfconsciously used to disprove my 'pessimism'. One man was convinced that if only the IPCC adopted a double-blind peer review system all of the criticisms of deniers would melt away. Another was convinced the climate problem will be solved through the development of a new energy source derived from high-flying kites, which could, he said, fully displace coal-fired electricity within a decade. An American woman accepted everything I said but simply evinced, with a shining face, an unbounded optimism that something would crop up. An ecologist argued that if we could put an economic value on ecosystem services, then the politicians would immediately understand why it is essential to protect the environment, although at the end of our conversation she mentioned that her three-year-old grandchild will probably be alive in 2100, at which point her eyes filled with tears of despair.

Some people derive a peculiar sort of pleasure in describing themselves as 'an optimist'. It's a kind of one-upmanship used to shut down those arguing that the evidence shows the future is not rosy. 'Whatever you might say, I am an optimist,' they declare, implying that their interlocutor is somehow not bold enough to take on the challenge. It's not so much passive aggression as sunny aggression firmly rooted in the moral superiority of cheerfulness, a modern predilection exposed by Barbara Ehrenreich in her excoriating book *Smile or Die: How Positive Thinking Fooled America and the World*. If positive thinking can defeat breast cancer, why can't it defeat climate change?

The power of wishful thinking, in which we allow our *hopes* for how things will turn out to override the *evidence* of how they will turn out, can be seen in some of history's great acts of unpreparedness. In 1933 Winston Churchill began warning of the belligerent intentions of Hitler's Germany and the threat it posed to world peace. In many speeches through the 1930s he devoted himself to alerting Britons to the dangerous currents running through Europe, returning over and over to the martial nature of the Nazi regime, the rapid rearming of Germany, and Britain's lack of readiness for hostilities.[112] Yet pacifist sentiment among the British public, still traumatized by the memory of the Great War, provided a white noise of wishful thinking that muffled the warnings. Behind the unwillingness to rearm and resist aggression lay the gulf between the future Britons hoped for – one of peace – and the future the evidence indicated was approaching: war in Europe; just as today behind the unwillingness to cut emissions lies the gulf between the future we hope for – continued stability and prosperity – and the future the evidence tells us is approaching: one of danger and sacrifice.

Throughout the 1930s Churchill's aim was, in the words of his biographer, 'to prick the bloated bladder of soggy hopes' for enduring peace.[113] But the bladder had a tough skin, far too tough to be penetrated by mere facts, even the 'great new fact' of German rearmament, which, said Churchill, 'throws almost all other issues into the background'.[114] The warnings of Churchill and a handful of others were met with derision. In terms akin to those now used in the Murdoch media to ridicule individuals warning of climate disaster – 'fear-mongers', 'doom-sayers', 'alarmists' – he was repeatedly accused of exaggerating the danger, of irresponsibility, of using 'the language of blind and causeless panic' and of behaving like 'a Malay running amok'.[115]

Late in 1938, Churchill's trenchant criticism of Chamberlain's Munich agreement – he attacked it as 'a total and unmitigated defeat' – earned him the fury of Conservative Party members. Anti-Churchill forces in the party rallied and as late as March 1939 – months before war was declared and a year before he was to become wartime prime minister – it seemed likely Churchill would be ousted as a Conservative MP by government loyalists. Although the evidence of German rearmament was incontrovertible, no one could be certain of the Third Reich's intentions, a doubt that provided the leverage for Churchill's critics. Of course, they all melted into historical obscurity when war broke out, which makes one wonder how, some decades hence, we will look on today's climate deniers whose campaign slowed efforts to protect the Earth from the ravages of climate change. Perhaps some of them have an intimation of the historical retribution coming their way and are taking out an insurance policy by supporting geoengineering.

5

Promethean Dreams

Fine-tuning the climate

Everyone is looking for an easy way out. The easiest way out is to refuse to accept there is a predicament. Another is to hope that the problem is not as bad as it seems and that something will come along. The technofix of geoengineering is a third way out and an emerging lobby group of scientists, investors and political actors is giving it momentum. Yet the appeal of climate engineering runs deeper, for as an answer to global warming it dovetails perfectly with the modernist urge to exert control over nature by technological means, a predisposition we begin to explore in this chapter.

Scientists, entrepreneurs and generals have long dreamed of controlling the weather.[1] The development of computers and the accumulation of weather data using satellites have prompted a new and higher phase of dreaming. In 2002 the American Meteorological Society published a NASA-funded study titled 'Controlling the global weather'. The author, Ross Hoffman, foresees the creation of an international weather control agency within the next three to four decades. 'Just imagine,' he enthused, 'no droughts, no tornadoes, no snowstorms during rush hour etc.'[2] Control would be possible, the argument goes, precisely because weather systems are chaotic. Chaotic systems are very sensitive to small perturbations, so if we can identify and then control those perturbations then we can control the weather: 'since small differences in initial conditions can grow exponentially, small but correctly chosen perturbations induce large changes in the evolution of the simulated weather'.[3] He did not dwell on the implications of small but incorrectly chosen perturbations.

Controlling one country's weather is not possible without affecting that of others, so the only way forward would be a global weather control system. Without close collaboration, Hoffman warns, there may be 'weather wars'. Among the perturbations that could serve as control mechanisms for global weather he identifies the timing and location of aircraft contrails, solar reflectors that regulate the amount of sunlight and an enormous grid of fans that could redirect atmospheric momentum.[4] A more recent scientific paper explores the possibility of a control strategy for El Niño, the periodic warming of central and eastern Pacific currents that causes drought in Australia and floods in South America.[5] It too looks for leverage in small disturbances with large effects, the most promising lever being alteration of sea surface temperatures in the eastern Pacific through cloud brightening.

Stephen Salter, an engineer and principal researcher in marine cloud brightening, is convinced that we will soon know everything there is to know: 'Noise is only a signal which you have not learned to decode yet.'[6] He is excited by the prospect of total control of the Earth's climate, and entertains plans of domination that would do Dr Strangelove proud. He defends further research with the claim that:

> we might discover that to get more rain at Timbuktu in August
> but less rain during Wimbledon you should spray to the west of

Cape Verde island from mid April to mid May and stop all spraying south of Kergulen during January and February. However spraying south of Tasmania from June to December never affects anywhere north of Hong Kong. By linking the strength of the beneficial effects with observations of the weather patterns and spray planning we may eventually develop sufficient understanding to allow tactical or closed-loop control which could respond to other more random perturbing influences and make everyone happier with their weather.[7]

This kind of technological hubris, although not often expressed so brazenly in public, colours the advice governments are beginning to receive from geoengineering researchers. The idea is taking root that geoengineering could be used not just to counter some of the effects of global warming, but to manipulate permanently the planet's weather system to suit our desires, or at least the desires of those who turn the knobs. To this end, climate engineers are beginning to talk about employing not one but a suite of interventions designed to tailor the climate. So stratospheric aerosol spraying might be used to cool the globe overall, while cloud seeding may be undertaken to fine-tune other environmental goals, such as preserving coral reefs, 'hurricane emasculation' and restoring polar ice caps.[8] Engineering the global climate thus becomes an optimization problem.

No two researchers are more prone to the special kind of scientific excitement that can possess geoengineers than Ken Caldeira and Lowell Wood. We saw that damage to the ozone layer is likely to be one of the side effects of sulphate aerosol spraying, allowing more ultraviolet light to reach the surface, so risking more skin cancer. Caldeira and Wood have an answer. They argue that some kinds of ultraviolet light that cannot be seen 'may be largely superfluous . . . for biospheric purposes, and thus portions of these spectra may be attractive candidates for being scattered back into space by an engineered scattering system.'[9] This light is invisible to us, so why do we need it? Particles could be specially engineered to allow through more of some kinds of light than others. They argue that such a scheme could save us $10 billion a year from avoided skin cancers. An additional benefit of scattering redundant bands of the light spectrum is that the sky could be rendered discernibly bluer.

It is a strange kind of thinking that believes it can identify basic properties of the solar system that are surplus to requirements and may be dispensed with. A different kind of thinking assumes that things are there for a purpose and that the structure of life on Earth as a whole has evolved to fit the environment in which it finds itself. So on closer inspection 'junk DNA' turns out to be genetic material whose functions we had not yet worked out. Many insects rely on ultraviolet light for their vision, reptiles need it to bask in and it is essential to production of vitamin D. The multitudes of species on Earth have evolved to manage the potential damage from ultraviolet light. Yet Caldeira and Wood suggest that we can filter out this superfluous form of light, so that we regulate not only the quantity of light reaching the planet but its quality. There is no bridge to cross to engage with this type of thinking. There is only an abyss of incomprehension.

The Promethean plan for ultimate control has been set out explicitly by Brad Allenby, now an engineering professor at Arizona State University, in a strategy he calls earth system engineering and management.[10] He begins with the observation that humans have not merely transformed the landscape but have imprinted themselves on every cubic metre of air and water, to the point where the Earth has become a human artefact. There is no more 'natural' so we must cast off all romantic notions and take responsibility for conscious planetary management. In a definition whose training manual phraseology says as much about its meaning as the words themselves, Allenby writes:

> Earth systems engineering and management may be defined as the capability to rationally engineer and manage human technology systems and related elements of natural systems in such a way as to provide the *requisite functionality* while facilitating the active management of strongly coupled natural systems.[11]

In case it might be thought that such a vision excludes all that is essentially human, Dr Allenby (who for some years in the 1990s was director for Energy and Environmental Systems at Lawrence Livermore National Laboratory) assures us that ethics can be incorporated into his system. It can even encompass 'religion', while still maintaining the requisite functionality, thereby granting space for a system-compatible God. To reassure those who fear that managing the Earth system must entail 'centralized control' or 'universal mandates', Allenby is certain that engineering an artificial world can be carried out by the free market. Moreover, he writes, Earth system engineering will embody 'inclusive dialog among all stakeholders' and 'democratic governance', while at the same time being modelled on 'highly reliable organizations' such as a well-run nuclear power plant or an aircraft carrier.[12]

It's hard to know what to make of this kind of utopian techno-enthusiasm, except to note that it is very prevalent in the geoengineering community, especially in the United States. It drives Bill Gates, Richard Branson and Nathan Myhrvold and a hundred other techno-entrepreneurs whose understanding of the world has been shaped by the peculiar culture of Silicon Valley. Brad Allenby has more recently shifted his position, tempering his dream of Promethean mastery with a strong dose of political conservatism.[13] Now he argues that climate science is disputable (there is a 'real controversy' over whether warming is caused by human-induced emissions or changes in solar energy) and that climate scientists do not have the same authority as other scientists. He believes, following standard denialist tropes, that contrarians have been unfairly 'demonized' and political polarization is due, not to the efforts of the merchants of doubt such as ExxonMobil and the Tea Party, but to the 'strident tone' of environmentalists. International collaboration won't work, he believes, but there is little need for it because the prevailing social and economic systems are adapting to climate change (such as it is) 'remarkably quickly'. No major policy interventions are needed, and that goes for geoengineering too. In short, the system is flexible and its components can adapt to whatever the climate throws at us; the real danger lies in overreacting to the apparent threat. Allenby has joined the small but influential group of 'luke-warmists', those who cannot be accused of denying climate science but consistently emphasize the uncertainties, downplay the risks and defend the prevailing order against policies that seem to threaten it.[14]

Other experts with a more clear-eyed view of climate science and its implications are turning their attention to the kind of engineering system that would be needed for managing the solar filter. The Novim Group, a non-profit scientific corporation, identifies five core control variables available for the solar filter or 'short-wave climate engineering' (SWCE): the material composition of the aerosol particles, their size and shape, the amount dispersed, the location of dispersal into the stratosphere and the sequencing over time of the injections.[15]

The development of a dynamic multivariate control system –
incorporating robust monitoring of climate parameters,
maximum intervention flexibility and intervention stability – is
therefore an important component of SWCE research. Control-
system design should pay particular attention to the likelihood
of various climate parameter responses including delays, feedbacks, nonlinearities and instabilities across widely ranging
temporal and spatial scales.[16]

Temporarily forgetting just why they are detailing Plan B, the
authors add that 'strategic management' of greenhouse gas emissions 'must be considered a central component' in managing the
solar shield. Good luck with that.

The engineers are alert to the fact that installing a planetary
thermal control system is not merely a technical problem. They are
concerned that unspecified 'socio-political system failures' –
perhaps climate wars, terrorist attacks, changes of government in
the US and social unrest in China – may lead to 'unintentional
disengagement' giving rise to 'transient oscillations in the climate
system'.[17] Transient oscillations in the climate system may refer to
monsoon failure, but the climate engineers are not too worried
because 'disruptions of varying character and scale are common in
comparably large and complex technical and socio-political
systems'. What were they thinking of when referring to disruptions
to comparably large and complex socio-political systems – the
Russian Revolution, the Great Depression, the Black Death? Who
knows? Even so, any control-system blueprint, they advise, should
keep these possibilities firmly in mind.

The Novim experts then canvass the dystopian prospect of
'counter-climate engineering' – geoengineering deployed by one
nation to undo the effects of geoengineering by another. 'For

example, the deliberate injection of short-lived fluorocarbon greenhouse gases might rapidly offset the regional or global cooling effects of a SWCE intervention.'[18] (In the case of marine cloud brightening, the fleet of unmanned ships roaming the oceans would be sitting ducks for a disgruntled state.) Any such contest over global weather could be 'disastrous', so international governance arrangements should be carefully considered. They finish on an optimistic note, suggesting that 'once engaged, the maintenance of a SWCE system becomes a permanent bequest to future generations'. A bequest to future generations. Words sometimes fail.

Some of those environmentalists and scientists most acutely aware of the dangers of global warming support geoengineering. Humans have caused such a build-up of greenhouse gases in the atmosphere, they argue, that even radical cuts in global greenhouse gas emissions will not be enough.[19] To render the climate tolerably safe we will need to reduce atmospheric concentrations of carbon dioxide to 350 ppm or below from their expected peak at 450 ppm (an extremely optimistic target), 550 ppm (optimistic) or 650 ppm (likely on current trends), remembering that the long-term pre-industrial level was 280 ppm.[20] It's a powerful argument with the best motives. By endorsing geoengineering their objective is not to find a way of defending the political and economic systems from the threat of climate change, but simply to protect us from calamity.

With their high level of understanding of the complexities of the climate system and the risks of global warming, those who take this position tend to favour early deployment of geoengineering because, even with radical abatement measures, carbon dioxide 'drawdown' will be necessary. So the sooner we start deployment of carbon dioxide removal methods the better. They tend to prefer more natural and local kinds of climate engineering such as reforestation and biochar rather than system-altering approaches such as ocean fertilization or a solar filter. The former are slow-acting methods that would require decades to take full effect and would therefore be of no use as a response to a climate emergency.

The grander climate engineering proposals operate on a scale far larger than previous interventions by humans in environmental systems. Nevertheless, some lessons can be learned from prior attempts to manipulate environmental systems.[21] The history of human interventions in complex ecosystems shows that they frequently trigger a burst of unintended effects. In one case, a fresh-water shrimp was introduced into a Montana lake in order to augment the food supply of salmon. However, it was not understood that shrimp feed at night, while salmon feed during the day, so instead of the salmon eating the shrimp, the two species competed for the same zooplankton food source. Instead of salmon numbers multiplying they fell, and so did those of the local eagle population that depended on them for food, undoubtedly with flow-on effects elsewhere. The intervention was a kind of 'ecological roulette' – spin the wheel and see what happens.

Human interventions have had many successes,[22] but it's the disasters that we should heed when considering schemes as audacious as some of those proposed by geoengineers. Success depends above all on minimizing the chances of unintended consequences, which in turn depends in large measure on limiting the effects to a bounded geographical area. A disaster following an attempt to manipulate the Earth as a whole would render trivial those resulting from the introduction of the beetle-eating cane toad in Queensland and the rat-eating mongoose in Hawaii. In their review of the lessons of biological control, Damon Matthews and Sarah Turner write that this kind of miscalculation would be unlikely today because of our greater understanding of ecological processes, although they recognize that humans are entirely capable of repeating errors even when knowledge of the consequences is readily available. The assumption that humans learn from their blunders is rarely a safe one.

In trying to get a sense of the likelihood of unintended consequences from system-altering geoengineering schemes, the primary lesson from the study of biological interventions is that the risks increase with both the degree of system complexity and the limits to our understanding of those systems. To date, biological interventions have been confined to ecosystems that are bounded in various ways, so the damage is limited. In the case of system-altering climate engineering schemes the local is the global: every major and minor ecosystem process would be changed by sulphate aerosol injection, marine cloud brightening or ocean fertilization (just as it is by global warming). The complexity of the Earth system is almost inconceivably deep. Even with leaps in understanding over the next decades, a cascade of unanticipated consequences from intervention seems inevitable. And we return to the disconcerting fact that, despite the enormous advances in climate science over the last two to three decades, each advance opens up new areas of uncertainty. While advances in climate science ought to be teaching us to be more humble, advocates of schemes aimed at regulating sunlight or interfering in Earth-system processes seem to draw the opposite conclusion.

We know that ecosystems behave eccentrically, even ones artificially created for their simplicity. They change rapidly over short time-frames, and often develop over long time-frames in ways we barely understand. While Lowell Wood bullishly proclaims: 'We've engineered every other environment we live in – why not the planet?',[23] a more humble scientist, Ron Prinn, has asked: 'How can you engineer a system you don't understand?'[24]

Thermo-economics

Before the engineers are permitted to implement their climate control system, a decision must be made about where to set the thermostat. Here the kind of technocratic thinking of the climate engineers fits neatly with the calculative reckoning of a certain type of economics. A widely circulated paper on the economics of geoengineering by Eric Bickel and Lee Lane first identifies the positive and negative impacts of climate change, policies to reduce greenhouse gas emissions, and sulphate aerosol spraying. It then evaluates costs and benefits by placing dollar values on the various effects over time, before concluding that the benefits of geoengineering vastly outweigh the costs.[25] Using the DICE model of Yale economist William Nordhaus (DICE standing for Dynamic Integrated Model of Climate and the Economy), the authors calculate that every dollar spent on injecting sulphur dioxide into the stratosphere will generate $25 in returns, a rate of return that makes the hype of the Planktos expedition appear modest. In this way, they work out how to set an optimal temperature for the Earth for the next *two hundred years*. This optimal path specifies where the global thermostat should be set to balance out the costs of geoengineering with the damage caused by the increase in temperature. As figure 4 shows, Bickel and Lane have calculated that allowing the Earth to warm by 3.5°C is the 'optimal' amount after two centuries.[26]

It would be pointless to deconstruct the economic analysis.[27] One is reminded of Ralph Waldo Emerson's aphorism: 'Nature hates calculators.' Any reader who has reached this point in the book will now understand that multiplying the uncertainties about the impacts of global warming by the greater uncertainties about the effects of sulphate aerosol spraying, then converting the result

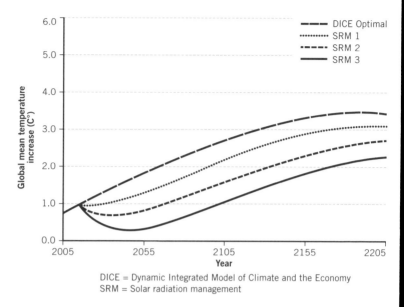

DICE = Dynamic Integrated Model of Climate and the Economy
SRM = Solar radiation management

Figure 4 Projected warming of the Earth through to 2205 under various solar radiation management regimes.

into net benefits by applying dollar values derived from economic forecasts 200 years into the future reaches a peak of analytical absurdity that is perhaps unsurpassed in the history of economics. When tax accountants are weighing up whether they should attempt to slip a questionable expenses claim past the tax office, they apply what they call the 'laugh test'. If there were a laugh test for economic modelling, Bickel and Lane would be doing stand-up at the Comedy Club.

So who are they? Lee Lane has been a 'resident scholar' at the American Enterprise Institute, a main cog in the denial machine, and is a consultant with CRA International, a consulting firm with a long history of working for the coal industry. Eric Bickel is an engineer at the University of Texas and a consultant to the oil and gas industry.[28] He appeared in *Cool It*, Bjorn Lomborg's film denigrating climate science. Bickel and Lane's paper on the economics of geoengineering originated at the American Enterprise Institute and was subsequently published by Lomborg's Copenhagen Consensus Center.

In their paper Bickel and Lane lament the fact that 'twenty years of protracted diplomatic talk and laborious scientific study have so far failed to move the needle on emission rates',[29] as if those who sponsor their work – the American Enterprise Institute and Lomborg's Center – played no role in preventing the needle from slowing. They refer back to an earlier faux-scholarly paper by Lane and David Montgomery, Vice President of CRA International, also published by the American Enterprise Institute, which argues that advocates of emission cuts fail to ask the 'most important and pressing questions', namely how it is that 'political structures can sometimes block or distort the choice of the best response to a problem'.[30] They blame the IPCC for the continued rise of emissions, suggesting that progress 'has been almost purely rhetorical'. Of course, the objective of the American Enterprise Institute is precisely to block or distort the best response to climate change. Straight-faced through 50 pages of argument, the authors evince a confected concern at the inability of governments to curb emissions growth, before expressing their gratitude to ExxonMobil for financial support.[31]

Bickel and Lane opine that one of the advantages of solar radiation management is that nations 'with relatively weak environmental lobbies' – meaning China and Russia – will be able to deploy solar radiation management without domestic opposition.[32] If they are right that it will not fly in a democracy, then solar radiation management is the dictator's technology of choice. What is disturbing is that Lane and Bickel count the ability to by-pass democracy as one of the *benefits* of solar radiation management as a response to climate change.

The political objective of the paper is transparent. The authors praise solar radiation management because, compared to measures to cut carbon emissions, it 'involves no infringement of economic freedom'.[33] They are not reluctant to declare that the value of technofixes is that they can solve a problem that would otherwise require social change. Geoengineering is an essentially conservative technology, one whose political appeal risks obscuring its inherent faults. This is why the economists who lend support to it – William Nordhaus, Thomas Schelling, Lawrence Summers, Richard Tol – are uniformly free market conservatives. In a deft reframing, Holly Buck has observed that the global energy regime needs to be changed, not least to provide opportunities in poor countries for development that is less dependent on fossil fuels. So to the extent that a solar filter would lock in existing social structures it involves a very large 'social opportunity cost'.[34] Of course, the economists whose mode of analysis is designed to defend the status quo do not accept that change is needed.

The Livermore taint

It is striking to realize how many scientists working on geoengineering have either worked at or collaborated with the Lawrence Livermore National Laboratory, the Cold War nuclear weapons facility. Among the prominent players I have mentioned or will mention, Edward Teller, Lowell Wood, Roderick Hyde, Ken Caldeira, Mike MacCracken, Greg Rau, Ron Lehman, Jane Long, Brad Allenby, Govindasamy Bala and Haroon Kheshgi have at some stage worked there. Although there are differences between them (MacCracken in particular was not part of the 'weapons crowd'), the common institutional origins and the early role of Teller and Wood have given geoengineering what Eli Kintisch calls 'the Livermore taint'.[35] This suggests that the history of Livermore may provide some insights into the deeper ideas behind climate engineering and the way some geoengineers understand the human relationship to the Earth. The best source on the scientific culture and political history of Lawrence Livermore National Laboratory is *Nuclear Rites: A Weapons Laboratory at the End of the Cold War* by anthropologist Hugh Gusterson.[36] I will suggest that the kind of intellectual culture that characterized weapons research at Livermore during the Cold War is having a formative influence on the geoengineering debate in the United States and beyond.

Along with Los Alamos, the Livermore Laboratory, located outside San Francisco, developed the hardware for the nuclear arms race; but it also influenced the Cold War ideology that went with it. The Laboratory was established in 1952 by a consortium of the University of California and a number of major weapons corporations, but most of its funding came from the federal government. The Laboratory was at the centre of the US programme to design a range of nuclear warheads and earned what Jeff Goodell calls a 'near-mythological status as the dark heart of weapons research'.[37] It was co-founded by Ernest Lawrence, who had received the Nobel Prize for physics, and Edward Teller, soon to become the most vigorous advocate of the hydrogen bomb (which in the late 1940s other atomic physicists had opposed as 'an evil thing considered in any light').[38] Teller was a Hungarian Jew who left Germany in 1933, arriving in the United States in 1935. Within a few years he became a central figure in the Manhattan Project, the Allied programme to develop an atomic bomb. After the war, as the actual or effective leader at Livermore, Teller was to become what one analyst described as 'a major architect of the cold war. . . .

[H]e did more than any other scientist, perhaps any other individual, to keep its structure intact and evolving'.[39] In the 1980s Teller was still regarded by some as 'the most powerful scientist in the world'. Although his influence waned, he remained 'the science hero of the Republican right'[40] and in the 1990s President Bush described him as 'my friend of long standing'.[41] Teller died in 2003, but his ghost haunts the debate over geoengineering today.

The influence in Washington of Teller and his protégé Lowell Wood caused ructions at Livermore itself. In the mid-1980s its director of nuclear weapons research, Roy Woodruff, claimed that Teller and Wood were providing misleading information to senior White House officials and arms control negotiators about progress in Star Wars technology, the nuclear-powered space shield championed by Teller. (Woodruff's intervention to correct the misrepresentation of the science is a useful reminder that physicists played a vital role in the criticism of Star Wars, and indeed the nuclear arms race more broadly.)[42] Although his claims were vindicated a few years later when Star Wars was abandoned, Woodruff was forbidden from correcting the misleading claims and resigned.[43] The incident illustrated both Teller's sway and the dangers of scientists becoming advocates for favoured technologies.

Weapons researchers came to believe that their technical expertise gave them a privileged role in advising government on nuclear policy. Washington concurred, going so far as to include Livermore scientists in the identification of nuclear targets in the Soviet Union, which is perhaps why the Russians called Livermore 'the City of Death'.[44] They also had a large role in deciding on the types of weapons to build. One said: 'if you don't understand the technology and physical effects of the weapons, then in my view you don't have the right to an opinion on nuclear policy'.[45] Among weapons scientists the conviction grew that understanding and exercising control of the technologies was sufficient to render them safe, as if mastery of the technical sphere carried over into the political sphere. Confidence in the technology spilled over into the structures that determined how and when it might be used, reflecting the modern predilection to elevate technical truths over other kinds of truths, so that those who could articulate the former acquired authority to speak.

In the emerging geoengineering field, scientists have assumed a privileged place in advising not merely on technical questions but on governance arrangements, ethical concerns and international negotiations, despite their lack of expertise. There is a view that if you are clever enough to understand atmospheric physics then you are clever enough to grasp the nuances of politics, social change and ethics. As in the nuclear arms race, the allocation of authority to those with scientific expertise reflects the continued privileging of the hyper-rationality of physical science over the kinds of reasoning and knowledge valid in other spheres where the weaknesses of humans and their institutions are recognized and the lessons of history absorbed.

Gusterson found, contrary to expectations, that weapons scientists at Livermore held a variety of political views, with as many identifying as liberal as conservative. They traversed a range of religious orientations: three even identified as Buddhists.[46] The emerging divide over geoengineering is not principally along a left–right fault-line, or even a pro-environment versus pro-economy split. So David Keith identifies as an environmentalist who loves the outdoors, and Ken Caldeira helped organize anti-nuclear protests as a student and describes his friend Lowell Wood as 'a right-wing nut'.[47] The divide is between Prometheans and Soterians, a technocratic rationalist worldview confident of humanity's ability to control nature, against a more humble outlook suspicious of unnatural technological solutions and the hubris of mastery projects.

Nuclear weapons scientists did not avoid ethical concerns about their work but learned to locate their activities within certain officially accepted frames. In the context of the Cold War and the argument for nuclear deterrence, it was not difficult. Civil protests caused them to reflect on their work but rarely to alter their views. Gusterson notes that, among the dozens he interviewed, they uniformly adopted a consequentialist ethical framework, that is, one in which moral judgements depend on weighing up the presumed benefits and harms to humans, as opposed to making judgements based on the virtue of motives, duty to a higher principle or anxiety about the morality of pursuing a good end by threatening to use unspeakably awful means.[48] From there it was an easy step to believing that developing nuclear weapons was *more* ethical than working on conventional weapons, because conventional weapons are used to kill while nuclear weapons exist to prevent killing. Some geoengineering scientists adopt the same kind of argument – attempting to develop means to prevent the worst ravages of a warming world is the most ethical action to take. It's a widespread and compelling view. For nuclear scientists the argument works only if deterrence is effective and nuclear weapons are never used. Once they had been used, in Japan in 1945, the only justification could be that their use saved more lives than the alternative course of action (even though in selecting Hiroshima from among several short-listed targets, one objective of military planners was to kill as many people as possible in order to demonstrate the power of the new weapon).

Livermore scientists were not opposed to nuclear arms control treaties, but they were 'almost unanimously hostile' towards test bans. There is a similarly strong resistance among geoengineers of the Promethean persuasion to any regulation of research and testing, especially from 'the UN'. At Livermore, antipathy to test bans was not merely pragmatic. Gusterson divined deeper cultural meaning in testing. The 'display of the secret knowledge's power' imparted a keen sense of community among participants. He read weapons tests as 'powerful rituals celebrating human command over the secrets of life and death'.[49] Tests were proof that human mastery of dangerous powers could be attained. In the same way we might expect that tests of geoengineering technologies, if they succeed, will persuade those carrying them out that technologies of planetary control can be mastered.

The analogy can be stretched too far. Although both groups of scientists are dominated by physicists and, in their different ways, both are developing technologies to control titanic powers, nuclear weapons scientists worked in tightly defined communities under conditions of secrecy in an officially sanctioned programme of defending the United States against an easily identified, human enemy. Geoengineering scientists do not set out to use technology to dominate an enemy but to respond to a non-human threat. Like nuclear weapons scientists they are competitive professionally, but unlike them they are spread across a range of institutions. Even so, the core group in the United States has developed a strong sense of group identity, apparent on internet discussion groups, that solidifies in response to outside criticism.

Gusterson notes that nuclear weapons scientists frequently used metaphors of birth to describe their work – generations of weapons, breeders, cribs and cradles. One described the feeling after a test as 'postpartum depression'. They named the Hiroshima bomb 'Little Boy' and after the first hydrogen bomb test Edward Teller telegrammed Los Alamos with the words 'It's a boy' to signal its success. A dud or 'fizzle' was known as a girl.[50] The language of climate engineering is evolving but already it is plain that attempts are being made to frame it semantically as something that works benignly with nature rather than against it, a technology that is remedial, even nurturing, rather than dominating and controlling: in short, a technology with feminine virtues.

The physicists who became weapons scientists were trained to believe that emotions are an obstacle to logical thought. At Livermore they derided the emotionalism of the anti-nuclear protesters who sometimes camped at the Laboratory's gates – 'a lot of hysteria but not a lot of solid thinking behind them', sniffed one. When asked whether he, like some Americans, ever had nightmares about nuclear war, a Livermore weaponeer responded: 'It's not rational to have nightmares about nuclear weapons. There's nothing you can do about them.'[51] For less rational beings, the terror lies in the helplessness.

Those who worked at Livermore found a culture in which brilliant and often quirky scientists dreamed up and tested big technological schemes to protect American freedom and advance US strategic interests around the world. In the 1980s the Reagan Administration poured billions of dollars into Livermore to fund the Star Wars programme that promised a fleet of nuclear-powered satellites that could use enormously powerful lasers to vaporize Russian missiles. At the heart of Star Wars were Teller and Wood.[52] Wood's 'personal fiefdom'[53] was known as O Group, which focused on how to use space for strategic advantage on Earth and included plans for an array of orbiting mirrors to train particle beams on enemy missiles. O Group also dreamed up Brilliant Pebbles, a fleet of orbiting space 'rocks' that could be deployed to collide with enemy missiles.

If Cold War thinking could be congealed into an institution, it would take the shape of the Lawrence Livermore National Laboratory. Yet in disarmament negotiations during the years of *perestroika* from 1985, reforming Soviet leader Mikhail Gorbachev would tell US presidents that all sides would need to move away from Cold War thinking.[54] In place of Gorbachev's more gradual process of democratic and market reform in the Soviet Union, Boris Yeltsin's machinations led to the disastrous 1991 collapse. Instead of Russians achieving their own freedom, anti-communists in the United States claimed victory. Conservatives began writing of how Ronald Reagan had won the Cold War. Lowell Wood himself spoke of how he was 'proud of the small part I played in this historic victory'.[55] Such an interpretation of history vindicated the kind of thinking that Livermore embodied. Technological supremacy, militarization of strategic policy and a unilateralist intolerance of collaborative agreements under the United Nations had prevailed. These same tendencies are emerging in thinking about climate engineering.

With the end of the arms race after the Soviet Union's collapse, Livermore, 'the most feared laboratory in the world', lost much of its *raison d'être*. While its leaders argued that weapons scientists were still needed to respond to threats from emergent nuclear nations and terrorist groups, they also began to look for new opportunities to keep Livermore relevant. As it happened, nuclear weapons research spilled over into atmospheric science. One of its tasks was to evaluate the effects of a nuclear exchange on the climate – nuclear winters – which required the development of sophisticated models to track the distribution of smoke, dust and radiation. This capacity was expanded in the 1990s to study the enhanced greenhouse effect.

In 1997 the Laboratory published a paper by Teller, Wood and Roderick Hyde, a senior scientist at Livermore, titled 'Global warming and ice ages: Prospects for physics-based modulation of global change'. Although not convinced that anthropogenic global warming was actually a problem, the authors nevertheless argued that any untoward changes in the Earth's climate may be better solved with technological interventions instead of 'international measures focused on prohibitions'.[56] Expressing cynicism towards democratic decision-making, they argued that a new technology of solar radiation management would be able to cut through international disagreements and win over public support. 'Physics-based modulation' would trump attempts at global consensus. In an article published in the *Wall Street Journal* two months later, Teller appealed both to economics and American exceptionalism. After noting that 'the jury is still out' on global warming, and deploring an 'all out war' on people who use fossil fuels, he wrote: 'Let's play to our uniquely American strengths in innovation and technology to offset any global warming by the least costly means possible.'[57]

Five years later, Teller, Wood and Hyde returned to the theme with a paper arguing even more strongly that the world should regulate solar radiation instead of attempting to reduce greenhouse gas emissions.[58] Although still casting doubt on the science of climate change, they set out ways to manage solar radiation actively (or practise 'radiative budget control', as they called it), which they insisted is the most practical approach to global warming. They drew on work by Ken Caldeira and Govindasamy Bala (also from Lawrence Livermore Laboratory) to claim that the environmental risks of a solar shield would be 'negligible'. The question, they wrote, is no longer whether we should engage in geoengineering but only how best to do it. And with the bombast typical of some Cold War physicists they insisted that those who accept climate science must '*necessarily* [their emphasis] prefer active technical management of radiation ... to administrative management of greenhouse gas inputs to the Earth's atmosphere'.[59] Not only is it technically easier, politically preferable and environmentally benign, the Livermore trio claimed, but sulphate aerosol spraying has large economic benefits – in the form of reduced skin cancer rates and improved agricultural productivity due to higher carbon dioxide – that render it worth undertaking even in the *absence* of any problem of global warming. We should be regulating the Earth's sunlight no matter what. Moreover, they went on, although the negotiators at the 1992 Rio convention did not realize it at the time, we may be legally obliged to install a solar shield because it is the cheapest option, so that under the UN Framework Convention on Climate Change geoengineering becomes *mandatory*.

Technically brilliant people can at times be very stupid and it is tempting to dismiss these claims as the fantasies of techno-maniacs trapped in the Cold War. Yet Western culture is often blind to the foolishness of genius, and the thinking of the Livermore trio is accorded a respect that is mystifying to those who can see its perils, not least many other geoengineering researchers who might share the Promethean ambition but are more careful to avoid being put in the 'mad scientist' box. Nevertheless, the appeal of these kinds of arguments has spread well beyond the laboratories. I have already mentioned the American Enterprise Institute and the Heartland Institute. The Hoover Institution, a conservative think tank partly funded by ExxonMobil and with a long association with climate science denial, has also promoted geoengineering.[60] Teller, until his death in 2003, and Wood have been Hoover fellows. So it should be no surprise that the US military is taking an interest in climate engineering.

Creeping militarization

In October 2011 a report strongly backing research into geoengineering was released by the Bipartisan Policy Center,[61] a non-profit

Washington think tank that claims to be politically balanced, although it's not clear what a middle position can be if one party repudiates science and the other does not – half the science? Investigative journalist Robert Dreyfuss has described the Center as 'a collection of neoconservatives, hawks, and neoliberal interventionists'.[62] The Center's engagement in geoengineering is significant in itself because it means that the issue has moved 'inside the Beltway'. While the subject was off the table only five years ago, the normalization of geoengineering as a legitimate response to global warming is now proceeding rapidly. The next IPCC report will take a giant stride in that direction.

The Bipartisan Policy Center taskforce that prepared the report decided to rebrand geoengineering as 'climate remediation', which it deemed a more 'useful' term. The term, which has been likened to 'clean coal', equates installation of a planetary solar shield with, say, decontaminating a former industrial site, a stretch by any measure.

In addition to the ubiquitous Ken Caldeira, David Keith and Granger Morgan,[63] taskforce members included David Whelan, Boeing's defence systems chief who for many years worked on weapons projects in the Pentagon's research arm, the Defense Advanced Research Projects Agency (DARPA). It also drew in Frank Loy, chief climate negotiator in the Clinton Administration, Ron Lehman, director of the Center for Global Security Research at the Lawrence Livermore National Laboratory, and Daniel Sarewitz, an author of the luke-warmist Hartwell report.[64] It was chaired by Jane Long, associate director-at-large at the Lawrence Livermore National Laboratory, and Stephen Rademaker, a former Assistant Secretary of State and now a Washington corporate lobbyist.[65] The taskforce was described by journalist John Vidal as 'the cream of the emerging science and military-led geoengineering lobby'.[66]

The panel had counted among its number ethicist Stephen Gardiner, whose work on climate ethics reflects a Soterian predisposition, but Gardiner withdrew because he felt that the process did not take the ethics of geoengineering seriously. He objected to the rebranding of geoengineering, saying that it cannot be regarded as a 'remedy' for anything. And he was uncomfortable with the 'coalition of the willing' approach to research and possible deployment because it was really a coalition of the willing and able.[67]

The report made the mandatory statements about mitigation being preferred, called for a major research programme and opposed deployment until it was shown to be worthwhile. More interesting was the explicit attempt to locate decision-making over geoengineering in the White House, that is, at the highest level of governance. Between the lines of the report was an argument for the United States to seize the initiative, with a warning that research activities are already underway in Germany, India, Russia and the United Kingdom. The taskforce wants to resist international treaties, repeating the standard conservative criticisms of the 'suboptimal' decision by the Convention on Biodiversity to regulate ocean fertilization – premature regulation will stifle research and will not prevent unilateral deployment. It prefers oversight by a coalition of nations that can be trusted under a set of informal 'norms'. As in most of the expert inquiries into geoengineering, despite the routine nod towards 'consultation' and 'transparency' (although the Bipartisan Policy Center forgot to disclose the financial interests of some taskforce members), the sentiment of the report is essentially anti-democratic. The more astute geoengineers are alive to the danger that nothing would kill off their research more quickly than the impression of a secret programme controlled by a powerful few, so democracy is embraced instrumentally, as a process that has to be negotiated in order to achieve their goals.

Alarm bells should have rung in 2009 when the Pentagon's DARPA convened a meeting to consider geoengineering.[68] Along with the involvement of the Lawrence Livermore National Laboratory, it was sign of the perhaps inevitable militarization of climate engineering. After all, officer training includes study of the often decisive role of the weather in battles. DARPA's mission is 'to maintain the technological superiority of the US military and prevent technological surprise from harming our national security. We also create technological surprise for our adversaries . . . .' The semi-secret, military-linked JASON group of top scientists is also reported to be studying geoengineering.[69] In 2011 the RAND Corporation, a think tank with deep links to the US military, published a 'political and technical vulnerability analysis' of geoengineering options as part of its National Defense Research Initiative.[70] Pointing out the troubling uncertainties inherent in climate manipulation, it encouraged the US government to establish international norms to govern geoengineering research. (These anxieties do not bother Lowell Wood, who has said there is no point arguing with him about geoengineering because deployment is 'written in the stars'.)[71]

Those familiar with the history of the US military's close engagement with major technological programmes, including those with no obvious defence value, are not surprised by creeping militarization of geoengineering. Historian James Fleming refers to the 'long paper trail of climate and weather modification studies by the Pentagon and other government agencies',[72] and concludes that 'geoscientists with high-level security clearances share associations, values, and interests with national security elites'.[73] Military planners have for decades imagined 'weaponizing' the weather; it formed an important focus of research on both sides of the Cold War. A 1996 paper commissioned by the US Air Force anticipated that by 2025 weather modification would be 'a force multiplier with tremendous power that could be exploited across the full spectrum of war-fighting environments' and that the United States could not afford to allow an adversary to obtain exclusive capability over the technologies.[74] It concluded with the declaration: 'We can own the weather.' Climate change itself is now seen by the Pentagon not only as a force multiplier but as a conflict multiplier. A 2003 study it commissioned urged consideration of geoengineering options to control the climate.[75]

With studies showing that a solar filter would have uneven regional impacts, the 'optimal' deployment strategy will depend on where you live.[76] In such a situation, consensus becomes more elusive and the powerful are inclined to assert their will. Divergences in regional impacts are expected to grow over time under solar radiation management, making international agreement increasingly difficult and raising the potential for conflict. In the end decisions about climate control would become expressions of economic and military power. As Fleming observes after reviewing a century of climate wizardry: 'If, as history shows, fantasies of weather and climate control have chiefly served commercial and military interests, why should we expect the future to be different?'[77]

Europe versus America

In the 1960s and 1970s, the kind of scientific thinking that characterized the military-industrial complex of the Cold War era faced a powerful challenge from 'environmental impact science'.[78] Prompted by Rachel Carson's *Silent Spring*, the book that initiated the modern environment movement, independent scientists began scrutinizing the damage to natural systems and human health from chemicals used in a variety of industries. The hegemony of science that uncritically served industrial expansion was met with research that questioned the unalloyed benefits of 'progress'. The threat to the established view that civilization rests on the base of science helps to explain the sustained fury of the backlash we see today. The muscular conservative belief that Western civilization rests on its superiority in science went into retreat but is now advancing again. It has been resurrected, for example, by Niall Ferguson in *Civilization: The West and the Rest*.[79] The revolt against environmental impact science, in which climate science has become embroiled, can be understood as a campaign to protect the values, institutions and privileges that sustained the scientific-technological order of the Cold War era.[80]

It is the desire to see Prometheus returned to the throne, from where he was unseated by Soteria, that explains why conservatives can both repudiate climate science, surely the pinnacle of Enlightenment rationality, and support geoengineering. Reflecting the human propensity to admit an ailment only if the medicine is palatable, some supporters of geoengineering regard it as an affir-mation of a natural order in which technologically advanced humans exercise mastery over nature, a direct repudiation of the environmentalist narrative that overambitious attempts to domi-nate nature are bound to come to grief.

In 2000 David Keith wrote that, although it had waned in the 1980s and 1990s, interest in geoengineering would return because 'the drive to impose human rationality on the disorder of nature by technological means constitutes a central element of the modernist program'.[81] In this he is surely correct. Yet Keith lamented only the absence of debate about the appropriate extent of planetary management and looked forward to the day when the Earth's climate and carbon dioxide concentration 'are seen as elements of the earth system to be actively managed'. He endorsed the view expressed in 1977 by the National Academy of Sciences when it urged Americans to ask: 'What *should* the atmospheric carbon dioxide content be over the next century or two to achieve an optimum global climate?'[82]

Faith in technofixes and an unapologetic desire to make money out of climate engineering bring a distinctively American world-view to climate engineering. There is of course a long history of criticism of technofixes in the United States, yet implicit faith in humanity's ability to overcome threats and master the environment defined the science-as-saviour culture of that nation in the post-war decades. Like their counterparts in Europe, American geoengineers are worried about the threat of global warming; but they are less likely to accept any intrinsic reason why, if we have the means, we would not take control of the planet as a whole. It is true that it was a Dutch scientist working in Germany who let the geoengineering cat out of the bag, but only after intense soul-searching. It is also true that the godfather of climate engineering was Edward Teller, a naturalized American who grew up in Germany, which only confirms the penchant of converts to be more Catholic than the Pope.

While acknowledging exceptions on both sides, I have noticed a marked difference in attitude among geoengineering scientists in the United States and Europe. The Promethean ambition of planetary control – perhaps expressed most starkly by Lowell Wood when he declared, 'We've engineered every other environment we live in, why not the planet?' – is harder to defend in Europe. In Western Europe, and especially Germany, geoengineering is regarded with more circumspection and anxiety. The complexity and capriciousness of the Earth are accorded a greater respect, and there is a historical reservoir of mistrust for the good intentions of humans intoxicated with technological power. If in broad terms the

American approach is Promethean, the European approach is
Soterian. The latter characterization, as we will see, is not so easily
applied to Russia.

It should not be thought that those more anxious about global
warming are more likely to support climate engineering, just as it
should not be thought that those who deny global warming science
reject it. Table 2 attempts to locate some of the individuals and
organizations I mention on a grid with two axes – concern about
climate change (alarmed, hopeful or not too worried, in denial)

Table 2 Anxiety about global warming and support for geoengineering

Strongly supportive of climate engineering	Mike MacCracken, Greg Rau, Stephen Salter, *Climate Code Red*	Lowell Wood, Bjorn Lomborg, Haroon Kheshgi, Scott Barrett, *Superfreakonomics*, Brad Allenby (2000–2001)	Edward Teller, American Enterprise Institute, Newt Gingrich, Heartland Institute, Bickel and Lane, Yuri Izrael
Cautiously supportive	Paul Crutzen, Ken Caldeira, Granger Morgan, Bill Gates, Royal Society, IPCC?	David Keith, Breakthrough Institute, Bipartisan Policy Center, Hartwell Group	
Deeply sceptical or opposed	ETC Group, Stephen Gardiner, Alan Robock	Governments, Most NGOs, Brad Allenby (2012)	
	Alarmed about global warming	Hopeful or not too worried	In denial

and attitude to the development of geoengineering methods (opposed or deeply sceptical, cautiously supportive, strongly supportive). Of course, the classification conceals differences, including more nuanced positions regarding different types of climate intervention: it's easier to support research into biochar than sulphate aerosol spraying; most environmental groups are very worried but remain hopeful (publicly at least) that the situation can be rescued; and the ETC Group is opposed in principle, while Gardiner's scepticism grows from concern about the physical and social risks. The schema is instructive nevertheless.

In general, the disposition shifts from the Soterian to the Promethean as we move from the lower left-hand box to the upper right-hand box. There is a reason for this: the two dimensions are expressions of one underlying belief about the distribution of power. For those in the top right box, humans have the power. If anything threatens to knock humans from their perch as lords of the Earth then it is a problem easily solved with a greater assertion of human power. For those in the lower left box, human exertion of power over nature has frequently failed and attempts to rescue the situation with further exertions of power are bound to fail too.

While some may find the division between Promethean and Soterian worldviews a helpful means of understanding attitudes to geoengineering in the United States and Western Europe, those categories perhaps have less application elsewhere. Sorting out national positions on climate engineering is impractical at this embryonic stage in the debate, although some signs are evident. They hint at how the geopolitics of climate engineering may play out, and it is to this question we now turn.

6

Atmospheric Geopolitics

The Russian Prometheus

Military interest in weather control reached its zenith during the Cold War. In 1957 the chair of the MIT meteorology department wrote that human progress had been due largely to our ability to control the environment and that future progress could be had 'by taking the offensive through control of weather', adding that 'I shudder to think of the consequences of a prior Russian discovery of a feasible method for weather control.'[1] The technological hubris of American science in the Cold War was mirrored in the Soviet Union. Competition was intense not only for military supremacy and in the 'space race' but also in programmes for weather modification. Between the great rivals, the Promethean impetus was indistinguishable. In 1960, in a book titled *Man versus Climate*, two Russian meteorologists wrote: 'Today we are merely on the threshold of the conquest of nature. But if . . . the reader is convinced that man can really be the master of this planet and that the future is in his hands, then the authors will consider that they have fulfilled their purpose.'[2]

At around the same time, echoing his sworn enemies, Edward Teller declared: 'We will change the earth's surface to suit us.'[3] The declaration was made in the context of his plan, launched at the Lawrence Livermore National Laboratory, to use nuclear explosions to move mountains, open up canals and gouge out new ports.[4] He named it Project Plowshare, although there was no talk of reassigning nuclear swords. Soviet scientists too proposed reshaping the landscape by detonating nuclear blasts to re-route Siberian rivers. The technological breakthroughs of the Cold War seemed to open up the possibility of total human mastery over any natural obstacle, and it was the scientists who possessed these god-like powers.

Today, the world's most aggressive geoengineer is a Russian. Yuri Izrael is the first to carry out tests of aerosol spraying, albeit at low altitudes from helicopters, to gather data on the optical characteristics of various particles.[5] As director of the Research Institute of Global Climate and Ecology at the Russian Academy of Sciences in Moscow, Izrael was said to be close to President Putin. However, there is no evidence I am aware of that the Russian government views geoengineering with anything other than indifference, and it is probably unaware of it. Although an IPCC vice-chair until 2008, Izrael often gives the impression of being a climate science denier. He invited the guru of American deniers Richard Lindzen to a conference of the Russian Academy of Sciences in 2004, and he himself featured at a conference of the Heartland Institute in 2008. Izrael has variously claimed that warming will not be harmful, that the Kyoto Protocol has 'no scientific basis', and that it would be cheaper to resettle Bangladeshis threatened by sea-level rise. And he argues for geoengineering instead of emission cuts.[6] His critics in Russia describe him as a 'fossil communist'.[7]

If Izrael does not reject climate science wholesale, his frequent collaborator Andrei Illarionov, said to have been Putin's top economic adviser, does not hold back. Invoking nationalist resentments, he described the Kyoto Protocol as 'war' against Russia.[8] In the *Moscow Times* he wrote that it was killing off the world economy like an 'international Auschwitz'. In the London *Financial Times* he compared it to fascism and communism because it was 'an attack on basic human freedoms behind a smokescreen of propaganda'.[9] If Izrael is a fossilized communist, Illarionov is a newly minted libertarian, close enough to the ideology of the right-wing Cato Institute in Washington to be made a fellow. In Izrael and Illarionov, old Soviet communism and new Russian libertarianism come together to reject climate science and endorse climate engineering, proving once again that environmentalism has always been equally threatening to the established order in the Soviet Union and the United States.

There are some more reasonable Russian voices talking about geoengineering, including a handful of scientists modelling the impacts of sulphate aerosol spraying.[10] They take a realistic view of the prospects of mitigation, noting that 'CO_2 emissions increase is exceeding even the most pessimistic of the IPCC projections' and draw the conclusion that engineering the climate is inevitable: 'Therefore, humankind will be forced to apply geoengineering to counter the unwanted consequences of global warming.'[11] The results of their model show that sulphate aerosol spraying is most effective at reducing global temperature if it occurs near the equator and at an altitude of 22–24 kilometres, but it is the following kind of conclusion that is more likely to influence the attitude of the Russian government when the time comes to make decisions: 'A decrease in precipitation of about 10% can be expected in the middle and high latitudes of the Northern Hemisphere, particularly in most of Russia'.[12] They are alert to the dangers to the ozone layer from sulphate aerosol spraying, making the novel claim that climate engineering of this kind should not be deployed until the ozone hole has been repaired: 'emissions of ozone-destroying substances into the atmosphere must be significantly reduced in accordance with the Montreal Protocol by the time of the possible geoengineering forcing.'[13]

Tempting China

Although much of the analysis so far has focused on the United States, where most of the research and advocacy has occurred, it is likely that the future of geoengineering lies as much in China. In an important and timely analysis of the status of solar radiation management in China, sinologists Kingsley Edney and Jonathan Symons begin by reminding us that China is the world's biggest emitter of greenhouse gases and that its government is acutely aware of the harm that climate change is likely to do to the nation.[14] Both of these facts help to explain the effort China is putting into promoting the growth of low-carbon energy sources, while it continues to build coal-fired power plants. China today operates on an unstated compact between the Communist Party and the populace: the party commits to providing jobs and rising incomes and in return expects the people to accept its grip on power. This explains the paramount importance placed on maintaining high rates of economic growth; yet that same preoccupation has made China the world's biggest emitter and is the cause of the most anxiety among climate scientists, and anyone with a modicum of foresight. And the impact of China's helter-skelter growth on air pollution, water supplies and food safety is creating its own upsurge of social protest. A 2011 Gallup poll found that 57 per cent of Chinese people believe that environmental protection should take precedence 'even at the risk of curbing economic growth'. Only 21 per cent put more growth above the quality of the environment.[15]

After a thorough review of commentary in the media, Edney and Symons conclude that the openness of discussion indicates that the government has yet to adopt an official position on solar radiation management. Chinese scientists are alive to the emerging debate in the West. 'However,' they write, 'we have been informed by distinguished scientists in China that SRM is not currently being researched or developed locally.'[16] More recently, the Chinese government listed geoengineering among its geoscience research priorities. Scientists quoted in the Chinese media adopt a range of views, from scepticism about its effectiveness to support for research, although overall they view it negatively. One story published by the official news agency Xinhua is titled 'Experts pour cold water on "geoengineering", cite high level of difficulty in project implementation.'[17] A senior meteorologist, Zhu Congwen, stressed the unintended consequences and the puniess of humans in the face of nature in the past: 'we emphasized humanity's triumph over nature, which left us with some bitter lessons.' Given the preference of US economists for technofixes that obviate the need for social change, it is remarkable to read one Chinese scholar, Professor Tan Zhemin of Nanjing University, declaring that instead of resorting to technology China would be better off 'changing people's lifestyles'.[18]

Edney and Symons found that there is plenty of information available on geoengineering in China, with several lengthy feature articles discussing the pros and cons. Some headlines give a flavour of the recent coverage: 'How to save the warming earth by raising a "giant umbrella"'; 'Seeking a climate change "Plan B"'; 'Geoengineering technology: Method of last resort'; and '"Geoengineering", can it save the world?'. The coverage has moved from treating geoengineering as science fiction, still a prevalent trope in the West, to more serious consideration. Xinhua published an article in 2010 dwelling on some of the crucial political questions raised by geoengineering, including the risk of undermining abatement efforts and the danger of Western countries controlling the technology. (One scientist returned from China saying that after a severe cold spell in 2008 some people there were saying that the Americans had already started their geoengineering.)[19]

China is a big country with a variety of climatic zones and already the differential impacts of climate change may be making themselves felt, with severe drying in the north and more flooding in the south. Studies to date indicate that solar radiation management can lower the average temperature of the Earth but will have a range of effects at the regional level that are hard to predict. A vigorous internal debate would precede any deployment of a solar filter by China, for there are likely to be sharply differing views among provincial governments. China has a long history of cloud seeding, although it should not be assumed that support for local weather modification implies support for geoengineering. There are already provincial tensions over cloud seeding, with complaints by some provinces that others are stealing their rain.[20]

Over the last decade China has assumed a vital role in international climate change negotiations, and will take a close interest in climate engineering. Indeed, Edney and Symons expect that as the impacts of a changing climate become more severe, China will one day advocate deployment of a solar filter.[21] Worries about the global political fall-out might be overwhelmed by the need to maintain domestic stability and continued growth, even though the regional impacts are likely to be highly uncertain and a solar filter cannot be a permanent solution. Edney and Symons take the view that preserving global harmony may see China prefer that sulphate aerosol spraying be managed by an international body. It would vigorously resist any attempt at unilateral deployment by another power or a compact among Western nations, which would be seen as an attack on China's sovereignty. The response might include organizing the G77 group of developing nations to oppose this new form of 'ecological imperialism', although China's leadership of G77 was damaged at the Copenhagen conference in 2009.[22]

Interestingly, Edney and Symons speculate that the United States, G77 and China may align in resisting attempts by the European Union to restrict solar radiation management.[23] A China–US compact is on the cards, but it is not clear that the divisions that marked international negotiations over emission reductions will map onto attitudes to solar radiation management. It is true that a reluctance to adopt emission abatement measures makes solar radiation management appear more attractive, but we are already seeing trenchant opposition to climate manipulation from the South. There is plausibility in Edney and Symons's suggestion that nations may divide along pro-growth and pro-environment worldviews, but there are dangers in imposing on developing countries a Western-sourced divide between Prometheans and Soterians.

Even so, some commentators have found a similar cultural split in China. In the 1950s China scholar Joseph Needham made comment on the age-old competition between Taoist and Confucian schools of hydraulic engineering. Practising *wu wei* ('No action contrary to Nature'), the Taoist engineers 'believed that the great river [the Yellow River] should be given plenty of room to take whatever course it wanted', but 'those who believed primarily in low dykes set far apart were opposed by those who believed in the main strength of high and mighty dykes, set near together'.[24] If the Three Gorges Dam is emblematic, then today the Confucian engineers rule supreme, but the complex cultural history of China urges caution. Confucianism is said by some scholars to emphasize harmony, respect for cosmic totality and belief in the vitalism of nature,[25] which ought to place it in the Soterian camp. Be that as it may, neither Taoist nor Confucian wisdom is a match for the power of the individualism and materialism that have now overwhelmed China.

Regulating the engineers

Around the world, experts, legislatures and international bodies concerned at the 'wild west' character of geoengineering research have begun to consider how it might be regulated, and how climate intervention should be governed if at some future point deployment is proposed. While there is no sense of urgency about the need to put in place regulatory mechanisms, there is recognition that controls will be needed and that it is better to begin consideration of governance mechanisms sooner rather than later.

To this end various processes have been initiated aimed at developing proposals for the governance of research, testing and deployment. They include: an inquiry by the United Kingdom's House of Commons; an inquiry by the US Congress's Government Accountability Office; two reports by the Royal Society; a decision to include assessment of geoengineering in the IPCC's Fifth Assessment Report; and a resolution by parties to the Convention on Biodiversity seeking to limit testing of certain geoengineering technologies. Mention should also be made of the Asilomar conference, a private meeting of geoengineering researchers and other interested parties in 2010, which developed a set of voluntary guidelines for geoengineering researchers.[26]

Should regulation occur at a national or international level? The answer is not straightforward. Land-based carbon dioxide removal activities – air capture machines, biochar, reforestation – are likely to be regarded as a component of the host nation's national or international commitment to reduce its net greenhouse gas emissions. Domestic laws would regulate them, although to meet international obligations and to qualify for carbon credits they would become enmeshed in the existing international system of greenhouse gas regulation.

Where carbon dioxide removal activities occur in global commons, such as ocean fertilization, existing international treaties come into play. A number of treaties already impose a duty on parties not to cause significant transboundary harm.[27] The London Convention and London Protocol, designed to outlaw the dumping of wastes at sea, adopted a resolution in 2008 declaring that its provisions covered ocean fertilization, although it permitted legitimate scientific research while urging 'utmost caution'.

Compared with carbon dioxide removal, certain features of solar radiation management make governance more necessary and more difficult. Transboundary effects create the potential for conflict, arguably even more than greenhouse gas emissions themselves as geoengineering involves *deliberate* attempts at climate modification. Testing is likely to require new legal instruments and organizations. There is widespread concern that a unilateral attempt to reduce warming through sulphate aerosol injections is a possibility because the technology is feasible, effective and cheap. Any medium-sized or large country could deploy a fleet of modified aircraft with the aim of transforming the Earth's atmosphere.

While some conceive of governance options on a scale ranging from a total ban to an unregulated free-for-all, a more helpful, and less tendentious, conception is to consider options ranging from early, comprehensive and escalating regulation to late, minimal and fixed regulation. Despite the framing by some geoengineering researchers, regulation does not equate to 'bans' but means a system of oversight and assessment on a case-by-case basis. In most countries, any publicly funded research is already subject to various systems of approval and reporting.

Some scientists engaged in geoengineering research are fearful that any kind of regulation would inhibit their ability to pursue their research, a sentiment apparent at the Asilomar conference and in the Royal Society inquiry into governance of solar radiation management. To ward off interference they argue for 'bottom-up' regulation according to a set of professional norms or voluntary guidelines, and invoke the 'basic principle of scientific freedom', interpreted to mean that as long as they do not impose harm or physical risk on others, scientists should be free to conduct the research they choose.[28] This begs the questions of who should decide whether a piece of scientific research is potentially harmful.

For those who invoke the principle of scientific freedom, the only consideration in developing regulations governing research is the potential physical effects of experiments, so that research that has no or negligible physical impacts on the environment should not be subject to regulation. A more plausible view is that, while research may have no or negligible *physical* impacts on the environment, there are *social* implications of any research into climate engineering which provide a justification for regulation. There is legitimate concern that research and testing of solar radiation management technologies would set the world on a slippery slope to ultimate deployment. I have already described the formation of a constituency of scientists and investors whose interests lie in further research and deployment. Path-dependency and technological lock-in are well-recognized problems.[29]

Most people are alive to the danger that even talking about geoengineering will further delay mitigation, which explains why geoengineering was not on the table for serious discussion until 2006 when Paul Crutzen made his landmark intervention.

However, governments remain wary of any discussion of climate manipulation. A good illustration is the response in 2010 of the UK government to a report on the regulation of geoengineering developed by the House of Commons Science and Technology Committee.[30] The committee urged the government to promote discussions within the UN towards the development of an international regulatory framework. Reflecting its nervousness that discussing governance processes would appear implicitly to endorse geoengineering, the government took the view that such moves would be 'premature' and that, as a great deal of preparation would be needed, a regulatory framework is many years away. At the same time, it reiterated its position that mitigation should be the priority.

Even so, governments will be compelled to grapple with governance sooner or later. Let me outline four options. First, a collection of national governance regimes, with more or less coordination between states, could be left to emerge. Existing environmental laws in various jurisdictions cover various aspects of geoengineering research and testing, including: laws on ecosystem disturbance; laws covering air; land and water pollution, including toxic substances; and, perhaps, laws relating to weather modification. In addition, although no nation has developed specific regulatory structures, some forms of geoengineering research are subject to public oversight through funding bodies and the ethics procedures of universities and government agencies. There are inevitable gaps and ambiguities in existing laws. The 2009 LohaFex iron fertilization voyage went ahead despite a dispute over its legality.[31] The experiment was approved by the German research ministry but opposed by the German environment ministry, which believed it was contrary to the resolution of the Convention on Biological Diversity (CBD).

Second, a collection of codes of conduct, principles and soft law could be developed by the scientific community and applied to its activities. Two codes – the Oxford Principles and the Asilomar Recommendations – emphasizing disclosure, independent technical assessment and some kind of government oversight have been developed. While appealing in the early stages of research, such voluntary codes are unlikely to satisfy public demand as geoengineering research develops and real-world testing is planned, as we have already seen in Britain with the cancellation of the SPICE experiment.

Third, one or more existing international treaties or organizations could be co-opted or extended to encompass regulation of climate engineering. While a number of existing treaties have a bearing on geoengineering, none of them has the scope or capacity to provide a comprehensive regulatory regime, which is not surprising given the novelty and diversity of geoengineering methods. A 2012 study commissioned by the CBD assessed the state of the regulatory environment and noted that, in addition to the Convention itself, a number of international treaties have a bearing.[32] They include the Law of the Sea, the London Convention and London Protocol on ocean dumping, the Montreal Protocol on ozone protection, Space law and the Convention on Long Range Transboundary Air Pollution. The ENMOD Convention – whose full name is the Convention on the Prohibition of Military or Any Hostile Use of Environmental Modification Techniques – is also salient. Following their use in the Vietnam War, any deployment of weather modification methods for hostile purposes is outlawed. Nevertheless, for regulating climate engineering there is a big gap in existing international law. There is no law, for example, that could prevent an individual deploying a solar shield through sulphate aerosol spraying.

Fourth, a new treaty or international organization with the mandate of governing one or more category of geoengineering could be developed. Reflecting the early stage of the debates, the diversity of technologies, and the existing complexity of international law, there are no proposals for a comprehensive treaty governing geoengineering research.[33] As the momentum for climate engineering gathers pace, it is inevitable that UN agencies, and the Security Council, will become engaged.

Already at international forums, the putting of resolutions has forced national governments to articulate an early position. Perhaps the best indication of emerging official attitudes to climate engineering can be deduced from the positions parties took at the 2008 and 2010 conferences of the Convention on Biological Diversity. The 2008 conference passed a resolution placing a moratorium on all ocean fertilization experiments, and the 2010 conference adopted a broader resolution declaring that 'no climate-related geo-engineering activities that may affect biodiversity take place' without adequate scientific assessment of the risks, 'with the exception of small scale scientific research studies that would be conducted in a controlled setting' and even then 'subject to a thorough prior assessment of the potential impacts on the environment'. At these meetings, resistance to geoengineering was led by countries of the South, some of which expressed passionate opposition. The most prominent were the Philippines, Malaysia, Ecuador, Bolivia, South Africa, Ghana, Malawi and Ethiopia. They were supported by Sweden and Norway. India too spoke strongly in favour of the resolutions. Their objections were rooted not in fears that the big polluting nations would attempt to use geoengineering to shirk their responsibility to cut emissions but that messing with nature would be wrong and dangerous.[34]

Russia and Japan argued for weaker resolutions, and Canada put up some resistance, but overall there was no strong opposition to the restrictions on geoengineering research contained in the resolutions.[35] It's worth remembering that, although it generally adheres to its provisions, the United States is not a signatory to the CBD and so was not present. There is a view in the South that the momentum for geoengineering is coming from the United States and the United Kingdom, although the United Kingdom has taken a constructive stance at CBD meetings.

However, we are at the stage in the evolution of geoengineering where governments are concerned principally with the optics of their positions, where bold pledges come at no cost. Even Germany, which has been bolder than most, could declare that the CBD resolution was 'not binding' when the research ministry decided to approve the LohaFex iron fertilization experiment. And when circumstances become worrying enough to trigger serious plans for climate regulation, the major powers are not going to be too concerned about a non-binding resolution of the Convention on Biodiversity. Still, the CBD resolutions may be seen as the first well-aimed shots fired by elements of civil society in a long campaign. The communiqué agreed at the Rio+20 convention in 2012 expressed concern about ocean fertilization, resolving to treat it with 'utmost caution'.

Talk of regulation meets stiff resistance from the small group of scientists who carry out most of the research and dominate the various inquiries and meetings on the topic. It is likely that, over time, pressure will increase to make governance of all research, testing and deployment of solar radiation management subject to international decision-making, probably via an existing or new UN treaty, although Catherine Redgwell, an international legal scholar, believes it is inconceivable that there would be the political will for a new international instrument to regulate geoengineering.[36] A New Zealand legal scholar, Karen Scott, sees geoengineering as 'the next great challenge for international environmental law' and argues persuasively that the best solution would be the adoption of a geoengineering protocol to the United Nations Framework Convention on Climate Change.[37] The 1992 Convention, ratified by virtually all nations, provides the global framework for climate change negotiations, and could be modified to set out general principles for all forms of climate engineering. If intervention is authorized in principle, detailed regulation could be devolved to other international bodies. A third scholar, William Burns, argues that sulphate aerosol spraying could not be deployed because it violates intergenerational equity, 'a binding principle of international law' that is 'incorporated in a wide array of treaties, domestic and international case law, domestic law, and soft law instruments'.[38] In the midst of a global climate emergency, however, the niceties of international law may be expendable.

If fear of unilateral action in the future is ever present, measures to reduce the chances could be implemented now. As a model, the Convention on Biological Weapons has developed a number of so-called confidence-building measures 'to prevent or reduce the occurrence of ambiguities, doubts and suspicions and in order to improve international co-operation in the field of peaceful biological activities'.[39] An international agreement governing solar radiation management might begin by asking national governments to promote international collaboration among research teams, including opening a register of research activities, disseminating information on all research and testing, reporting annually to a UN body on all aspects of solar radiation management research and testing and facilitating inspection of research and testing facilities.

Desperate times

Any suspicion that one powerful nation is developing plans for unilateral deployment of solar radiation management would create fear and acrimony, rendering international collaboration both more necessary and more fraught. Already concern about unilateral deployment by a 'rogue state' has entered the public domain.[40] Sulphate aerosol spraying is cheap and technologically unsophisticated, and any number of nations could undertake it alone. Indeed, a billionaire with a messiah complex could do it. Perhaps in the face of a severe and prolonged drought or a series of scorching summers, the most likely candidates for unilateral deployment of solar radiation management are China and Russia. The scenario in which China decides to go it alone in the teeth of international protest is less likely than one in which China deploys a solar barrier with the tacit approval of the US government, which could fend off domestic protests with a display of faux indignation. The prospect of rapid deployment of a solar shield in the face of a 'climate emergency' has begun to attract serious consideration.

In 2006 NASA and the Carnegie Institution convened a workshop on 'managing solar radiation'.[41] It was the workshop for which Ken Caldeira coined the term 'solar radiation management' to soothe bureaucratic nerves over 'geoengineering'. It was attended by the main players in the world of geoengineering, including Lowell Wood, Haroon Kheshgi, David Keith, Scott Barrett, Phil Rasch, John Latham, Stephen Salter and Lee Lane.[42] The authors of the workshop's report were Caldeira and Lane, along with two scientists at NASA's Ames Research Center.

The report noted that the justifications for research into solar radiation management are built on two broad scenarios that could motivate deployment.[43] One 'strategic vision' sees it deployed pre-emptively once its feasibility and cost-effectiveness have been established. The soft version of this strategy sees solar radiation management as a technological intervention that can 'buy time' for the world to develop cheaper emission abatement technologies in the face of political difficulties in reaching international agreement. The hard version, pushed by free market economists, is that if solar radiation management is economically more efficient than reducing emissions then there is no reason why it should not be adopted as an alternative to abatement and deployed as soon as feasible.

The second strategic vision of the NASA workshop imagines rapid deployment of solar radiation management in response to a climate emergency. No definition of a climate emergency is given, but the Novim report has more recently obliged: 'We define *climate emergencies* as those circumstances where severe consequences of climate change occur too rapidly to be significantly averted by even immediate mitigation efforts'.[44] Actual or imminent sharp changes in world climate might include sudden and rapid melting of the Greenland ice sheet, acceleration of permafrost thawing or a prolonged and severe heatwave. The report goes on to note that different kinds of emergency would require different kinds of solar radiation management, with longer or shorter ramp-ups, and that some emergencies, such as the breaking off of the Ross Ice Shelf (Antarctica's biggest), could not be reversed by any kind intervention.

The NASA document notes that framing climate engineering as an emergency response effectively rules out carbon dioxide removal methods, as they take decades to have a substantial impact. This limits the discussion to stratospheric aerosol injections, and perhaps marine cloud brightening. Those who anticipate deployment in these conditions favour sustained research into solar radiation management so that, once developed and refined, the technology could be 'put on the shelf' to be used as necessary. 'In this situation,' the report suggests, 'politically, the decision to deploy solar radiation management would be relatively straightforward.'[45] The workshop participants were alert to the fact that their strategies have far-reaching implications for the governance and political legitimacy of solar radiation management. In words that would later be virtually cut and pasted into Lee Lane's report for the American Enterprise Institute, the authors argue that, in the emergency framing, there is no point thinking about political objections and popular resistance to solar radiation management because, in a crisis, 'ideological objections to solar radiation management may be swept aside.'[46] They count the ability to sweep aside civil society objections to deployment of solar radiation management as an 'obvious political advantage'. That the American Enterprise Institute should support the bypassing of democracy is no surprise; that a government agency should endorse such disdain for public participation in decisions determining the future of the planet comes as a shock.

A breakout group of the workshop proposed the development of a curriculum 'to train a generation of geo-engineers with emphasis on system engineering'. In another breakout group, on 'policy sciences', the assembled experts considered how to go about making solar radiation management acceptable to the public, suggesting that work could be done on understanding the political strategies of opponents and identifying international treaties that might act as barriers.[47] A pre-emptive deployment strategy 'could encounter strong resistance both domestically and abroad', but opinion research could reveal how proponents of solar radiation management can better use the threat of pre-emptive deployment as a 'bargaining chip'.[48]

Reading the report of the NASA workshop, it is hard to avoid the impression that this was a meeting of a highly select group of technical experts in positions of great influence who regard the democratic process as an obstacle to their plans. The disdain for public opinion in their own country, let alone the interests of the countless millions who may be harmed by deployment of a solar shield, oozes from the pages. They seem almost to welcome the arrival of a climate emergency in which democratic processes can be overridden by scientific experts who control the means to save the world.

Since the NASA meeting, arguments for research funding have increasingly relied on the prospect of a climate emergency.[49] Emergency responses often entail a decision by a powerful authority to override normal processes of democratic decision-making. Desperate times call for desperate measures. They usually give priority to one objective at the expense of others; in this case, solar radiation management intervention may be aimed at reducing warming, paying less attention to concerns about ocean acidification, drought in the Sahel or ozone depletion. Economic and natural emergencies are often exploited for political benefit, following the maxim 'never let a good crisis go to waste'. Those who defend solar radiation management research as a form of preparation for a crisis have yet to provide answers to the following questions: What are the criteria for a climate emergency requiring rapid intervention? Who would determine that an emergency exists? Who would authorize the emergency response, and from where would they derive their legitimacy? Who would decide that the emergency is over? Of course, the development of apparently effective solar radiation management technologies may well make a 'climate emergency' more likely; by reducing political and popular incentives to introduce abatement measures, the continued growth in emissions increases the likelihood of a crisis, at which point supporters of deployment will be able to claim that 'there is no alternative'.

All of these ideas about the geopolitics of climate engineering in a world under climate stress are only beginning to circulate in the opaque world of international strategizing. They give a flavour of the geopolitical imbroglios that geoengineering portends. It ought to be apparent by now that at the heart of geoengineering lie some deep ethical questions, and it is to these we now turn.

7

Ethical Anxieties

Justifying geo-research

At its core, climate engineering is a moral question. The same is true of all major environmental disputes – over nuclear power, genetically modified organisms and lead pollution. Each controversy has been driven by ethical arguments. In the case of geoengineering the moral landscape is only just beginning to be recognized.

In his 2006 intervention Paul Crutzen wrote: 'By far the preferred way to resolve the policy makers' dilemma is to lower the emissions of the greenhouse gases. However, so far, attempts in that direction have been grossly unsuccessful.'[1] The starting point for any consideration of the ethics of geoengineering is this failure of the world community to respond to the scientific warnings about the dangers of global warming by cutting greenhouse gas emissions. When I say it is a failure of the 'world community' this should not obscure the fact that it has been certain powerful nations, and certain powerful groups within those nations, that have been responsible for this failure.[2] As we will see, not all those promoting geoengineering research and deployment view it as a response to moral failure, so it helps to set out briefly the arguments in favour of research into geoengineering. I will refer mainly to the case of sulphate aerosol injections because it illustrates the ethical anxieties most starkly. Some of the arguments apply equally to the other system-altering technologies, notably ocean iron fertilization and marine cloud brightening, but may have less force when applied to more localized interventions.

Three main justifications are used to defend research into geoengineering and possible deployment: it will allow us to buy time, it will allow us to respond to a climate emergency and it may be the best option economically.

The *buying-time argument* – the main one used in favour of more research in the 2009 Royal Society report[3] – is based on an understanding of the failure to cut global emissions as arising either from political paralysis or from the power of vested interests. The logjam can only be broken by the development of a substantially cheaper alternative to fossil energy because countries will then adopt the new technologies for self-interested reasons. Sulphate aerosol spraying would allow warming to be controlled while this process unfolds. It is therefore a *necessary evil* deployed to head off a greater evil, the damage due to unchecked global warming. It is a powerful, pragmatic ethical argument for research into geoengineering and its possible deployment.

The *climate emergency argument* was Crutzen's motive for breaking the silence over geoengineering. Sulphate aerosol injection, he wrote, should only be developed 'to create a possibility to combat potentially drastic climate heating'.[4] Today it is an argument growing in influence, reflecting concern about climate tipping points.[5] It envisages rapid deployment of a solar filter in response to some actual or imminent abrupt change in world climate that cannot be averted even by the most determined mitigation effort. It's easy to see how, in some circumstances, this argument could be an overwhelming one. Although the solar shield may have draw-backs, failing to deploy it could result in much greater harm.

As in the case of buying time, the *best-option argument* would see sulphate aerosol spraying deployed pre-emptively rather than being left 'on the shelf' until an emergency occurs. Rejecting the understanding of geoengineering as an inferior response, it argues that there is nothing inherently good or bad in any approach to global warming. The decision rests on a comprehensive assess-ment of the consequences of each approach, which is often reduced to the economist's assessment of costs and benefits. In this narrow consequentialist or utilitarian approach the 'ethical' decision is the one that maximizes the ratio of benefits to costs. As we saw in chapter 5, some early economic modelling exercises have concluded that geoengineering is cheaper than mitigation and almost as effective and is therefore to be preferred.

Those economists who adopt the best-option argument see geoengineering as a potential substitute for mitigation rather than as a complement to it. They do not accept that geoengineering represents a necessary evil. If Plan B proves to be cheaper than Plan A then it would be unethical *not* to use it. They thereby avoid accusations that their advocacy undermines the incentive to choose the better path; geoengineering *is* the better path.

Philosophically, this requires the adoption of a narrow utili-tarian moral viewpoint. Using this framework it is possible to maintain that geoengineering is (or could be) a good thing only when all of the positive and negative effects of all plans are commensurable, so that one can be traded off against another. More precisely, for the utilitarian they can be traded off without feelings of guilt, regret or anguish.[6] We all at times have to make forced choices; but what makes some choices forced is that the decision entails a moral struggle. For the utilitarian no choices are forced because all effects can be traded off on rational grounds. Utilitarianism is the emotionless philosophy.

One implication of the narrow consequentialist approach is that there is nothing inherently preferable about the natural state, including the current climate. Depending on the assessment of human well-being, there may be a 'better' temperature or climate as a whole. In other words, it is ethically justified for humans to 'set the global thermostat' in their interests, if they can agree on what their interests are.

Another uncomfortable implication of the best-option argument is that it implies that the ethics of geoengineering actually excludes values because the argument can be reduced to apparently objective scientific and economic facts. So the utilitarian position rejects the view, implicit in the other two arguments, that motives count when making ethical judgements. The economists and utilitarian philosophers who adopt this approach see themselves as pragmatic – what matters, practically and ethically, is what works. Most people believe that intentions matter morally, which is why courts judge manslaughter less severely than murder. Against this everyday intuition, some philosophers argue that there is no defensible distinction between a harm caused intentionally and the same harm caused unintentionally, that the degree of 'wrongness' of an action has no bearing on its degree of 'badness'.[7]

The issue is complicated by the fact that, since we know that continuing to burn fossil fuels will cause harm, it could be said that global warming is now 'deliberate' even if warming is not the intention. Continued release of greenhouse gases is unquestionably negligent, but I think there is a moral, and certainly an attitudinal, leap to a conscious plan to modify the Earth's atmosphere. This is why studies concluding that sulphate aerosol spraying could disrupt the Indian monsoon are potentially explosive.[8] Certainly, one would expect the law to take the view that damage to someone arising from a deliberate action carries more culpability. In law, culpability for harms caused by an action depends in part on *mens rea*, literally 'guilty mind'.

Moral corruption

The psychological strategies we deploy to deny or, more commonly, to evade the facts of climate science, and thereby to blind ourselves to our moral responsibilities or reduce the pressure to act on them, were described in chapter 4. They include wishful thinking, blame-shifting and selective disengagement. For selfish reasons we do not want to change our behaviour or be required to do so by electing a government committed to deep cuts in emissions. Stephen Gardiner argues that this kind of situation gives rise to moral corruption, 'the subversion of our moral discourse to our own ends'.[9] Unlike the moral hazard and slippery slope arguments (considered in the next sections), moral corruption is not concerned with the consequences of our actions but with notions of 'bad faith', that is, duplicity and self-deception.

Climate engineering may lend itself to moral corruption. If we are preparing to pursue geoengineering for self-interested reasons – because we are unwilling to restructure our economies or adjust our lifestyles – then the promotion of geoengineering can provide us with a kind of cover or even self-absolution. But if climate engineering is inferior to cutting emissions (in the sense of being less effective and more risky) then merely by choosing to engineer the climate instead of cutting emissions we succumb to moral failure. It should be remembered that consideration of climate engineering comes after a long history of bad faith in international negotiations, in which various nations have engaged in pious declarations of concern coupled with ruthless obstructionism. Even when agreement has been reached, as in Kyoto in 1997, some nations have reneged on their commitments.

This picture of moral failure and bad faith cannot now be wiped away using some kind of historical Etch A Sketch, because climate engineering is a direct result of that failure. If we resort to climate engineering then the efforts of ExxonMobil, for example, to subvert the truth would be rewarded. More generally, those most negligent in carrying out their duties would be able to use geoengineering to avoid censure. Installing a solar filter would cement the failure of the North in its obligations to the global South. This is another way of making the case that what matters ethically about geoengineering is not only the outcome but also the human virtues or faults it reveals. In the end it might be judged that rewarding those guilty of bad faith is an unfortunate necessity.

Accusations of 'copping out' may not apply to those who are constrained in their actions, so that implementing the best plan is beyond their power. This is sometimes called the 'control condition' for moral responsibility.[10] It presents a moral dilemma for environmental groups: if they believe that Plan B is inferior to Plan A, then supporting geoengineering can be justified only if they believe they can no longer effectively advance Plan A. The dilemma deepens if it proves that supporting geoengineering actually makes emission reductions less likely to be pursued. Most environmental groups have adopted a wait-and-see approach, although they are instinctively suspicious of climate engineering.[11]

Scientists who defend geoengineering research are mostly exempt from the moral failings that have given rise to the situation. After all, most are among those who have supported strong abatement action and have become alarmed and frustrated at the failure of political leaders to act. It's not their fault and, deeply concerned, they are looking for ways of saving the world from the consequences of institutional failure. For both environmentalists and researchers who see geoengineering as a necessary evil, to maintain their integrity they must continue to argue that mitigation is superior. So, like Crutzen, the 2009 Royal Society report declared resolutely that mitigation is to be strongly preferred and geoengineering cannot be 'an easy or readily acceptable' alternative.[12]

Nevertheless, simply restating this belief may not be enough; unless one continues to act on it, the declaration risks becoming merely a means of deflecting censure. This draws attention to the position of governments and major fossil fuel corporations – for it would be hollow for them to argue that they are pursuing Plan B even though they believe Plan A is superior. They have the power to implement Plan A, or not to block it, and their reluctance or obstructiveness is the reason Plan B is being considered in the first place. To promote geoengineering they must convince others that it is not in their power to reduce emissions, a tactic that is frequently used. Even in the United States some argue that there is no point in cutting US emissions if other major emitters do not do the same, an appeal to the 'prisoner's dilemma' that all too easily serves as an excuse for inaction.

Recourse to the prisoner's dilemma – a situation in which it is in the collective interest to cooperate but, in the absence of trust, in the individual interest to behave selfishly and penalize others – is often an attempt to 'rationalize' moral decisions, that is, to shift surreptitiously to a consequentialist framing in which calculating outcomes gives the moral answer. Ethics becomes a 'game' that abolishes the motivation to do the right thing. So the prisoner's dilemma is often not a bona fide reason for lack of progress, but the opposite, a cover for bad faith. 'Hey, we want to do the right thing, but if others will not cooperate, what can we do?'

Moral corruption is a danger to geoengineering researchers tempted to accept financial support from governments, or fossil fuel corporations seeking to avoid their obligations. In 1962, noting the amount of money poured into universities by chemical companies, Rachel Carson observed: 'This situation ... explains the otherwise mystifying fact that certain outstanding entomologists are among the leading advocates of chemical control. Inquiry into the background of some of these men reveals that their entire research programme is supported by the chemical industry'.[13] Bad faith stains those who get too close. It's worth noting that when the time arrives at which they feel they can back research into climate engineering, governments and fossil fuel corporations are unlikely to appeal to the climate-emergency justification because highlighting the severity of global warming would only underline their moral failure. Moreover, as we saw, those able to implement emission cuts will lack credibility if they defend climate engineering with the buying-time argument. This leaves them with the best-option economic argument. In the case of the solar shield, the empirical basis remains speculative, not least because the risks of unintended consequences appear so high. Moreover, the appeal to economics as the basis for making such a momentous decision risks accusations of abandoning ethical concerns and treating the atmosphere as a resource.

The same moral failure arguments could be used by poor countries against rich ones. As it will probably be industrialized nations, including China, that invest most in geoengineering research and, if the time comes, deployment of the technologies that result, poor countries will accuse them of evading their obligations to reduce emissions. Studies indicating that some poor countries may suffer harms from some climate engineering techniques reinforce the likely sense of grievance. The ethical situation would be reversed if a small, poor and vulnerable country decided to protect itself by engineering the climate with sulphate aerosol spraying (something that may prove technically and financially feasible). The Maldives, for example, would have a strong moral case to argue that the threat to its citizens' survival from rising seas caused by the refusal of major emitting nations to change their ways, and its own inability to influence global warming despite sustained efforts, leave it with no choice.

Moral hazard

It is widely accepted that having more information is uniformly a good thing as it allows better decisions to be made. Geoengineering research is strongly defended on these grounds. Yet for many years research into geoengineering, and even public discussion of it, was frowned on by almost all climate scientists. As we saw, when Paul Crutzen made his famous intervention in 2006 calling for serious study of sulphate aerosol spraying he was heavily criticized by fellow scientists. They felt that investigating climate engineering would erode the incentive to reduce emissions, the response to global warming strongly preferred by scientists, including Crutzen himself.

In other words, they were worried about 'moral hazard', a concept developed by economists to capture the impact on incentives of being covered against losses. For example, it is argued that the incentive to drive a car carefully may be reduced if the driver is insured because the costs of an accident are spread across all who are insured. Although commonly used in the climate change context, the argument mistakenly transposes an understanding of incentives developed for private market behaviour into the realm of public policy decision-making. There are a number of ethical and practical objections to this move,[14] perhaps illustrated most starkly by the unwitting *reductio ad absurdum* embedded in the claim by economist Martin Weitzman that assessing the worth of 'life on Earth as we know it' is 'conceptually analogous' to deciding, for example, how much to pay for additional airbags in a car. Life on Earth itself is converted into a financial value by reducing it to how much we'd be willing to pay in the market.[15]

Nevertheless, the idea of moral hazard, suitably modified, is useful for drawing attention to political incentives. The availability of an inferior policy substitute that can be made to appear superior may make it easier for a government to act against the national interest.[16] It is well established that those whose financial interests would be damaged by abatement policies have been using their power in the political system to slow or prevent action.[17]

So does geoengineering research create moral hazard? Geoengineering researchers tend to be vague and somewhat dismissive of the likelihood, as though it is only of theoretical concern. The 2009 Royal Society report, dominated by geoengineering researchers, treats it as an uncertain effect that may even work the other way, and refers to some distinctly unpersuasive focus group work suggesting that individuals might increase their efforts to cut their emissions if government invested in geoengineering.[18] Overall, the report saw moral hazard (wrongly interpreted as concerning individual behaviour) as a 'factor to be taken into account', but in no way decisive.

Yet in practice any realistic assessment of how the world works must conclude that geoengineering research is virtually certain to reduce incentives to pursue emission reductions. This is apparent even now, before any substantial publicly funded research programmes have begun. Already a powerful predilection for finding excuses not to cut greenhouse gas emissions is obvious to all, so that any apparently plausible method of getting a party off the hook is likely to be seized upon. For the moment, governments and energy companies are staying at arm's length from geoengineering research, precisely because they fear being accused of wanting to evade their responsibilities. But the day when it becomes respectable to support geoengineering research cannot be far off. Already, representatives of the fossil fuel industry have begun to talk of geoengineering as a *substitute* for carbon abatement.[19] Economic analysis is in general not interested in the kind of judicious technology mix or emergency back-up defended by some scientists, but will readily conclude that geoengineering should be pursued, even as the sole solution, if that's what the cost curves show. Indeed, the popular but error-riddled book *Superfreakonomics* insists that the prospect of solar radiation management renders mitigation unnecessary: 'For anyone who loves cheap and simple solutions, things don't get much better.'[20] Instrumental thinking does not come much cruder, yet it is just this kind of Promethean wand-waving that prevails in the power centres of the world. For the authors, economics renders ethical concerns redundant: 'So once you eliminate the moralism and the angst, the task of reversing global warming boils down to a straightforward engineering problem: how to get thirty-four gallons per minute of sulfur dioxide into the stratosphere?'

We have seen that conservative think tanks are joining the fray, with the climate-denying Heartland Institute and American Enterprise Institute supporting climate engineering. Former Republican presidential candidate and House Speaker Newt Gingrich declared: 'Geoengineering holds forth the promise of addressing global warming concerns for just a few billion dollars a year. Instead of penalizing ordinary Americans, we would have an option to address global warming by rewarding scientific invention ... Bring on the American ingenuity. Stop the green pig.'[21] For these advocates the problem of moral hazard evaporates because there is nothing wrong with eroding the incentive to cut carbon emissions if a cheaper means of responding to global warming is available.

Gardiner has offered a left-field argument for the irrelevance of concerns about moral hazard.[22] After the Copenhagen fiasco, the prospects for substantial emissions abatement policies in the foreseeable future sank so low that the availability of a substitute to abatement could not drive them any lower. It is an argument from despair. Against it, in some parts of the world – notably the European Union and China – substantial efforts are being made to reduce emissions and accelerate the development of alternative energy technologies. In 2011 parliamentary support for the Australian government's carbon tax was on a knife-edge. Inadequate as they are, these efforts depend on a level of political resolve that could be weakened. Moreover, incentives to act could change rapidly as the effects of climate change become more obvious over the next decade and the availability of an apparently effective alternative to emission cuts could determine the kind of action taken.

That in practice moral hazard is the most powerful ethical argument against the development of geoengineering technologies is suggested by the highly germane case of carbon capture and storage (CCS).[23] Soon after the 1997 Kyoto agreement, the governments of the two nations that refused to ratify it, the United States and Australia, began talking up the benefits of CCS, a technology that promised to extract carbon dioxide from the smokestacks of coal-fired power plants, pipe it to suitable geological formations and bury it permanently deep beneath the earth. Burning coal would be rendered safe so there was no need to invite 'economic ruin' with policies mandating emission reductions. Quickly branded 'clean coal', the promise of the technology was increasingly relied on by the world coal industry to weaken policy commitments and spruce up its image.[24] The promise of CCS has been used repeatedly by both governments and industry as a justification for building new coal-fired power plants. In the United Kingdom, Prime Minister Gordon Brown declared that we must have it 'if we are to have any chance of meeting our global goals'.[25] US President Barack Obama's public endorsement of 'clean coal' was featured in PR videos made by the coal lobby.[26] German Chancellor Angela Merkel backed industry plans to build dozens of new coal-fired power plants, expecting that at some point they would be able to capture the carbon dioxide and send it to subterranean burial sites.[27] In Australia, the world's biggest coal exporter and the nation most dependent on coal for electricity, Prime Minister Kevin Rudd declared CCS 'critical' to generating jobs and bringing down greenhouse gas emissions.[28]

Economists also bet on the technological promise. The Stern report called CCS 'crucial'.[29] Jeffrey Sachs, Director of the Earth Institute, repeated the common opinion that there is no way China will stop building coal-fired power plants, so the technology 'had better work or we're in such a big mess we're not going to get out of it'.[30] The Garnaut report wrote that the success of 'clean coal' will ensure that any negative impacts of greenhouse policies on coal-dependent regions are 'many years away'.[31] The International Energy Agency promoted it enthusiastically, describing an ambitious roadmap for the deployment of the technology, to be led over the next decade by developed countries, after which 'CCS technology must also spread rapidly to the developing world', because without it costs of emissions reductions will be 70 per cent higher.[32]

Torrents of public funding flowed to CCS research. The Obama Administration's 2009 stimulus bill allocated US$3.4 billion and the US Department of Energy announced it would provide US\$2.4 billion to 'expand and accelerate the commercial deployment of carbon capture and storage technology'.[33] In the same month, the Rudd government in Australia announced it would commit A\$2.4 billion (around US\$2 billion at the time) to an industrial-scale demonstration project.[34] In 2009 the high hopes invested in CCS provoked the conservative business magazine *The Economist* to comment that 'the idea that clean coal . . . will save the world from global warming has become something of an article of faith among policymakers'.[35]

Yet from the outset impartial experts argued that the promise of CCS was exaggerated.[36] Even supporters of CCS conceded that the technology, if it worked, would have no impact on global emissions until at least the 2030s, well after the time scientists say deep emission cuts must begin. The most damning assessment was made in 2009 by *The Economist* in an editorial titled 'The illusion of clean coal':

> The world's leaders are counting on a fix for climate change that is at best uncertain and at worst unworkable. . . . CCS is not just a potential waste of money. It might also create a false sense of security about climate change, while depriving potentially cheaper methods of cutting emissions of cash and attention – all for the sake of placating the coal lobby.[37]

The Economist was echoing the warnings of critics who, from the outset, identified one of the major risks associated with pursuit of CCS as the way in which it would undermine global mitigation efforts by giving national governments an excuse to do nothing in the hope that coal plants could be rendered safe.

It turns out that the critics were right. Despite the hype, the hopes and the public investment, the promise of CCS is now collapsing. Its leading experts are expressing disappointment at the failure of governments and the coal industry to follow through on their commitments.[38] In November 2010 Shell's Barendrecht carbon capture project in the Netherlands was cancelled due to local opposition.[39] A month later ZeroGen, a huge project identified by the Australian government as a 'flagship' carbon capture project, was shelved because of cost blow-outs and technical difficulties.[40] The *New York Times* commented: 'Australia's experience with CCS mirrors technical, financial and political hurdles experienced in the United States.'[41]

There could not be a more vivid illustration of moral hazard than CCS, yet it is into this political and commercial environment that geoengineering arrives as the next great white hope. It is presented as a solution to the same global warming problem, to the same politicians, with the same recalcitrant industry, the same public prone to wishful thinking and the same largely uncritical media. The conditions are perfect for moral hazard.

The false promise of CCS played a vital role in the lost decade of response to climate change. Will geoengineering be the excuse for another lost decade? There is no sign that political leaders have been chastened by the sorry experience of CCS. If they are resolved to avoid difficult decisions and cosset the coal industry, why would they not just move on to the next technological boondoggle? Once the political threshold that currently restrains governments and coal companies from publicly backing geoengineering is crossed, warnings such as that made by the Royal Society – 'None of the methods evaluated in this study offer an immediate solution to the problem of climate change and it is unclear which, if any, may ever pass the tests required for potential deployment'[42] – are likely to be swamped by bold claims. The caveats at the front of geoengineering reports declaring that mitigation is the best solution will quietly disappear. Climate engineering has 'moral hazard' written all over it.

Lure of the technofix

If Plan B is inferior to Plan A the moral hazard is that its political attractions will undermine the incentive to cut emissions. So engineering the climate is deemed preferable despite the evidence. The slippery slope is an ethical concern closely related to moral hazard; moral hazard applies to policy-makers, while the slippery slope applies to those who back geoengineering. A lobby group of researchers, investors and, perhaps, regulators backing geoengineering is naturally inclined to overstate its benefits and understate its costs, and its risks. In 1962 Rachel Carson wrote: 'The chemical weed killers are a bright new toy. They work in a spectacular way; they give a giddy sense of power over nature to those who wield them, and as for the long-range and less obvious effects – these are easily brushed aside as the baseless imaginings of pessimists.'[43]

As we saw in chapter 4, the constituency for geoengineering is growing, and already reaches into stratospheric levels of wealth and power. Several influential inquiries have called for research programmes; it is being talked about in the White House; parliaments are interested; military planners are becoming engaged; venture capitalists and billionaires are investing; and patents are being registered. The conditions are ideal; we are already sliding down the slippery slope and it is only a matter of time before policy-makers come under intense pressure to choose the less desirable options. There is therefore a legitimate concern that the knowledge generated by geoengineering research will be misused in foreseeable ways.

However, the strength of the moral hazard and the slippery slope dangers depend in part on the absence of technological hurdles that appear insurmountable. While the experience with carbon capture and storage points to the strength of the moral hazard concern about geoengineering, it also suggests a brake on the slippery slope. A very powerful constituency formed around the promise of CCS, perhaps reaching its pinnacle with the creation in 2009 of the Global Carbon Capture and Storage Institute.[44] 'Clean coal' is not dead yet but, as the technical difficulties become more apparent, it is waning as a credible alternative to emission reductions, and the momentum is stalling. Yet a decade was lost. The slippery slope towards the deployment of, say, sulphate aerosol spraying will depend on continued research and testing not turning up some severe risk or insuperable obstacle that its more open-minded supporters cannot ignore. On the other hand, as the severity of global warming manifests, the penchant for downplaying the risks will intensify.

On the slippery slope, technologies gather added political momentum because we live in societies predisposed to seek technological answers to social problems. Previously, I have attributed our failure to cut emissions to political systems influenced by sectional interests, dominated by growth fetishism and led by individuals too timid to act on the scientific warnings. I have also attempted to explain widespread denial and evasion in terms of the comfortable conservatism of consumer society and the gradual alienation from nature.[45] Among conservatives there is a tendency to regard these as immutable facts of modern life. So instead of promoting change in political and social structures we are urged to resort to technological solutions that will bypass the blockages. Advocating far-reaching social change is dismissed as 'utopian'. But is social change in response to climate change impossible? Debating radical social change was part of the daily discourse of Western society from the time of the French Revolution until the 1980s, when the neoliberal revolution brought about 'the end of history', so the unwillingness today to consider changes to economic, social and political structures is all the more striking in the face of a threat as grave as the climate crisis. Shunning deeper questioning of the roots of the climate crisis avoids uncomfortable conclusions about social dysfunction and the need to challenge powerful interests. Calls for a technofix, including geoengineering, are thus deeply conformable with existing structures of power and a society based on continued consumerism. The slippery slope to the technofix promises a substitute for the slippery slope to 'revolution'.

An extreme kind of technofix as a response to global warming has recently been put forward by three philosophers (of the Anglo-American school) in a bizarre paper titled 'Human engineering and climate change'.[46] The authors, Liao, Sandberg and Roache, argue that we should consider seriously proposals to 'engineer' humans to reduce carbon emissions. One leading idea is genetic intervention to allow parents to select shorter children because smaller people eat less. They also use less petrol in their cars, need less energy-consuming fabric for their clothes, and wear out their shoes more slowly. If families had a cap on their emissions, parents could choose 'between two medium-sized children, or three small-sized children' or, if they wanted a basketball player, 'one really large child',[47] although an unintended consequence of using hormone treatment to create smaller children is a greater risk of gallstones.

Their other proposals for human engineering include geneti-cally engineering human eyes to be more like those of a cat because 'if everyone had cat eyes, you wouldn't need so much lighting'; reducing the birth rate by 'cognitively enhancing' unintelligent women because 'women with low cognitive ability are more likely to have children before age 18'; 'pharmacological enhancement of altruism and empathy'; and pills that make those who take them vomit if they eat beef, thereby reducing demand for beef.

The paper, published in a respectable journal, is beyond satire and its only likely effect is to bring the philosophy profession into disrepute. Analytical philosophy, it seems, does not have a 'laugh test' for filtering out whacky proposals. If we are to engineer humans to have cat's eyes and midget babies, why not genetically modify black people to make them white in order to cool the Earth by increasing its reflectivity?

Defending his decision to publish, the editor of the journal claims the authors are engaged in a 'Swiftian philosophical thought experiment'.[48] In fact, the opposite is true. Jonathan Swift's 'modest proposal' that poverty-stricken Irish peasants support themselves by selling their babies to be eaten by the rich – 'a young healthy child well nursed is at a year old a most delicious, nourishing, and wholesome food, whether stewed, roasted, baked, or boiled' – was a savage satire on the heartlessness of society in the face of mass suffering. The three philosophers are not lampooning our disregard of the threat of climate change. It is as if Swift had put forward his modest proposal as a legitimate response to famine. No doubt it could be wholly justified in utilitarian terms; indeed Swift himself carried out the cost-benefit analysis in order to heighten the ridicule.

The three bio-ethicists suggest that people who are appalled at the idea of human engineering may have a 'status quo bias', resisting their innovative ideas because of an unthinking conservatism. They seem oblivious to the irony, since their own proposal takes the technofix to a sublime plane, one made possible by an intensely individualistic understanding of the world, which sees the failure to respond to climate change as arising not from political, institutional and cultural forces but from a lack of personal willpower. Rarely in intellectual history has such a dire social problem been so trivialized by this kind of psychologism. The authors are keen to stress they would never *compel* people to produce small children or grow cat's eyes, which only raises the question of why anyone who is unwilling to buy a smaller car or switch to green power would be willing to genetically engineer their children.

In his critique of the Royal Society's 2009 report on geoengineering, Gardiner poses the question bluntly: 'if the problem is social and political, why isn't the solution social and political as well [and] if, as the report asserts, we already have adequate scientific and technological solutions, why assume that research on alternative solutions will help?'[49] In the end, the answer from geoengineering supporters must lie in an implicit judgement that social change is inconceivable so the only answer is to buy time for the costs of renewable energy technologies to fall far enough or to prepare to deal with an inevitable climate emergency. Yet in investing so much in our ability to take control of the climate are we in danger of attempting to emulate God?

Playing God

Weapons scientists inside the fence at the Lawrence Livermore National Laboratory were divided from the anti-nuclear protesters at the gates less by their political leanings or religious beliefs than by their commitment to a Promethean as opposed to a Soterian worldview. In a similar way, the concern about Promethean overreach, often known as 'playing God', is not confined to theists but may resonate just as strongly with atheists. For atheists, 'playing

God' is a metaphor either for humans assuming God-like attributes or for mortals attempting to occupy a domain that is not properly theirs. In the first, the idea is that there are certain qualities that humans cannot and should not aspire to, both because they are beyond us and because aspiring to them invites calamity.[50] The philosopher Tony Coady identifies three attributes of God or the godhead that are beyond human capabilities – omniscience, omnipotence and supreme benevolence – which seem to capture the common sense.[51] We will return to this meaning.

The second interpretation reflects a 'spatial' metaphysics of the world. Playing God entails humans crossing a boundary to a domain of control or causation that is beyond their rightful place. In this view, there is a limit to what humans should attempt or aspire to because the division between domains is part of the proper order of things. For theists, this other domain may be the dwelling place of God. For atheists, the domains are contained in an intuitive metaphysical order that defines 'the scheme of things' within which one can find what it means to be human. For both, the idea of staying out of 'God's realm' is an essentially Soterian outlook, sensitive to human shortcomings and the danger of ignoring them.

So what in practice is 'God's domain'? In the debate over human genetic enhancement the playing God argument has been prominent. Biologically, DNA is the essence of life, coding all of the information that makes an individual unique. As such, tinkering with genes (and especially the germ-line, or changes to DNA that can be passed on) can be seen by the theist as invading the sacred, or by the atheist as disturbing the essential dignity of the human. Michael Sandel argues that it is the gifted character of human capacities and potentialities that incites a natural reverence, and that there is something hubristic and unworthy about attempting to overrule or improve upon this gift through genetic enhancement. Manipulating genes to human ends is 'a Promethean aspiration to remake nature, including human nature, to serve our purposes and satisfy our desires'.[52] Life is reduced to a manipulable genetic code.[53]

The particulars are not of much help in the case of geoengineering because we are not talking about transforming humans but the world in which humans live. Yet global dimming via sulphate aerosol injections is a similarly Promethean aspiration to remake 'nature' to serve our purposes, this time not at the microscopic level of DNA but at the macroscopic level of the Earth as a whole. The domain being invaded is not that of the essential code of each life but the sphere in which all life was created or emerged. With solar radiation management the concern is not so much a lack of gratitude for a unique and precious gift, but the invasion of and dominion over the atmosphere that encompasses the planet – the benevolent ring that makes it habitable, supplies the air breathed by all living things and sends the weather. In most cultures for as long as humans have lived, the sky has been the Heavens, the dwelling place of the gods. Global dimming would not only transform the atmosphere but also regulate the light reaching the Earth from the Sun. For some cultures the Sun has its own divine character because it is the source of all growth, the food of plants and thus all living things. It is the origin of the most primordial rhythms that have always governed our lives – the cycles of day and night and the annual seasons. For those cultures the Sun *is* God, and attempting to regulate it would surely be out of bounds. I mention these cultural facts not because they *prove* anything but to invoke in the sympathetic reader a feeling for the role of the Sun as a symbol of powers beyond the reach of mortals. The popular preference, revealed by many surveys, for solar energy over nuclear power can probably be traced to a felt distinction between using a natural gift that flows freely to the Earth and relying on an unnatural and dangerous contrivance that has diabolical connotations. In general, people are more inclined to endorse technologies that appear to work with, rather than go against, nature.[54]

So the intuition is that the grander schemes to regulate the climate trespass in a domain properly beyond the human. To cross over successfully would require mortals to possess a degree of omniscience and omnipotence that has always been reserved for God or the great processes of Nature that are rightly beyond human interference. To make matters worse, in this view, we want to supplant the gods in order to counter the mess we have made as faulty humans. Instead of embarking on a vain quest to mimic the gods, it seems safer and more within our powers to face up to our failures and attempt to become better humans. The usual appeals to the power of reason and science make little headway because they are deployed in the service of the same conquering spirit that drives the desire to play God, as if human ratiocination can function as a battering ram to enter the gods' domain, there to dethrone them and elevate humans in their place.

So the first argument against mimicking God is not about the dire *consequences* of entering the domain of the gods, but that playing God betrays a deep fault in the human character. What of the second caution about playing God, that human aspirations to omniscience, omnipotence and benevolence invite calamity? In modern times, we have come to believe that the relentless accumulation of scientific knowledge is taking us closer to total understanding. Recent developments in Earth system science have increased our knowledge substantially, but they have also uncovered cavernous gaps. We have come to see more clearly that the climate system is extremely complex both in itself and because changes in it cannot be isolated from changes in the other elements of the Earth system. Human-induced warming is expected to reconfigure global precipitation patterns, but predictions of regional rainfall changes are very crude. The importance of tipping points' that define rapid shifts from one climate state to another have become apparent from the Earth's geological record, but our understanding of why and when they occur is rudimentary. Predicting when or how thresholds might be crossed is extremely imprecise. And how marine ecosystems will respond to acidifying oceans is barely grasped. In this light, *omniscience* appears as remote as ever.

Apart from the uncertainties, unknowns and threshold effects arising from the complexity and non-linearity of the Earth system, the dominant fact is that carbon dioxide persists in the atmosphere for many centuries. So it is possible – indeed, likely – that before the larger impacts of warming are felt, humans will have committed future generations to an irreversibly hostile climate lasting a thousand years. Yet some economists are telling us that they can use their models to estimate future streams of monetary costs and benefits to determine the optimal temperature of the Earth over the next two centuries, as if we know enough to install and begin to operate a 'global thermostat'. Truly this qualifies as monstrous hubris.

Humans are powerful, but what kind of power do we aspire to with climate engineering? Beyond deliberate management and exploitation of particular resources or geographical areas, and beyond the unintentional degradation of land, rivers and oceans, we now aspire to take control of and regulate the atmosphere and climate of the planet as a whole. As we will see in the next chapter, geoscientists are now arguing that humans have become a planetary force in their own right. We have so transformed the face of the

Earth that we have created a new geological epoch, one expected to be characterized by more climatic instability than the previous one. In other words, our Promethean aspirations have made the world *less* controllable.

If humans were sufficiently omniscient and omnipotent, would we, like God, use climate engineering methods *benevolently*? Earth system science cannot answer this question, but it hardly needs to, for we know the answer already. Given that humans are proposing to engineer the climate because of a cascade of institutional failings and self-interested behaviours, any suggestions that deployment of a solar shield would be done in a way that fulfilled the strongest principles of justice and compassion would lack credibility, to say the least. We find ourselves in a situation where geoengineering is being proposed because of our penchant for deceiving ourselves and inflating our virtues. If a just global warming solution cannot be found, who can believe in a just geoengineering regime? It is believed that a solar filter would offset some of the impacts of global warming more effectively in some parts of the world than others. In some areas it may even exacerbate droughts. The temptation of those who control the heat shield to manipulate it in a way that suits their interests first would be ever present and almost irresistible. And at no forum will non-human species have a voice. All of these anxieties are deepened by the creeping militarization of geoengineering and the possibility of unilateral deployment.

The playing God argument is not necessarily a categorical injunction against solar radiation management, but it does sound a warning about Promethean recklessness, calling for utmost caution and deep reflection. On one view, calculating risks is enough. On another, our attitudes and beliefs about ourselves and the nature of the world are so deeply ingrained that they necessarily constrain any calculative thinking to a narrow range of outcomes. According to this Soterian view, if we are so mistaken in our understanding of the world and our role in it that we are drawn into playing God with the future of the planet, then thinking must be grounded in a different relationship between humans and the natural world, one that recognizes the boundary between the domain of mortals and that of the gods.

8

This Goodly Frame

this goodly frame, the earth, seems to me a sterile promon-
tory . . . What a piece of work is a man!

Hamlet

At the outset of this book I described how interest in climate engineering blossomed with rising anxiety among climate scientists
about the widening gap between the actions demanded by the
evidence of global warming and the measures nations seemed
willing to take. The angst is due to a simple but paramount fact that
few among the public and political leaders have yet grasped – the
carbon dioxide we are putting into the atmosphere will persist,
altering the climate of the Earth, for thousands of years. This verity
makes climate change unlike any other environmental problem,
whose harms and solutions are marked by their immediacy. When
one's eyes are opened to the paramount fact, it is natural to shift
one's perspective, to stand back and rethink climate change and
geoengineering in the wide sweep of human and geological history.
In this last chapter I make some first, tentative observations from
this standpoint.

A new epoch dawns

The Earth's climate over the most recent 10,000 years has been remarkably stable. These ten millennia, the geological epoch known as the Holocene, were preceded by more than 100,000 years of climatic chaos. For much of it most of northern Eurasia and North America were buried under ice sheets several kilometres thick, yet the variability was enormous. Palaeoclimatologist William Burroughs writes that the climate 'swung from the depths of glacial frigidity to relative mildness [often] in the space of a few years'.[1]

> Armadas of icebergs or floods of icy fresh-water swept out into the North Atlantic altering the circulation of the ocean at a stroke and with it the climate of neighbouring continents. With a flick of the climate switch, Europe and much of North America could be plunged back into icy conditions, having only just emerged from the abyss of the preceding millennia.[2]

So for most of its 200,000 years on Earth, Homo sapiens has had to contend with a climate much more capricious and less friendly. Although it was overall much colder than today (especially in higher latitudes), over the last ice age sudden warming events occurred frequently. They often began with an abrupt warming of Greenland, with 5–10°C of warming over a few decades or less.[3] As ice sheets retreated, forests spread north from southern Europe as far as southern Russia. Cold snaps were just as frequent.

Although humans survived these wild swings, at times large populations were decimated. For instance, there was a collapse in the numbers of humans around 73,000 years ago, perhaps to as few as 10,000, a catastrophe thought to have followed the eruption of the super-volcano Mount Toba in Sumatra which blanketed South Asia with 15 centimetres of ash. The consequent 'volcanic winter' cooled the Earth by as much as 3–5°C for six to ten years and may have accelerated a glaciation, helping to cool the Earth further for a thousand years.

As can be seen in figure 5, around 20,000 years ago, following 100,000 years of chaotic climate, the Earth began a dramatic warming. Although interrupted 13,000 years ago by the Younger Dryas, a cold phase lasting a millennium, around 10,000 years ago the climate stabilized around an average temperature very close to the modern one prior to the influence of industrialization. The Holocene epoch had arrived. Rising sea levels were one of the consequences of the warming that gave us the Holocene, although

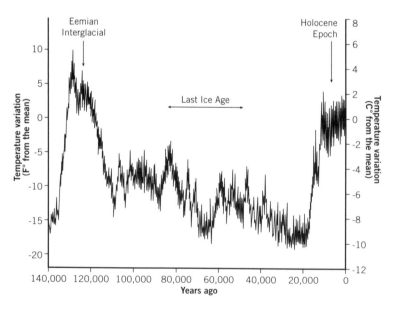

Figure 5 Variations (from recent average) in Antarctic temperatures over the last 140,000 years.

higher seas can lag many centuries behind a spell of warming. Twenty thousand years ago, sea levels were 130 metres lower than today.[4] Ten thousand years ago humans could still walk from England to France and from New Guinea to Australia.[5] The Holocene's temperature has been exceptionally constant, varying no more than half a degree on average, although more in some regions. The so-called Medieval Warm Period, which peaked in the tenth and eleventh centuries, and the Little Ice Age of the sixteenth to the nineteenth centuries were relatively minor and localized blips confined to parts of the northern hemisphere. Burroughs estimates that with the advent of the Holocene the variability in temperature declined by a factor of five to ten;[6] it was like walking out of the long grass on to a well-tended lawn.

It was the Holocene's unprecedented climatic stability and mildness that permitted human civilization to flourish. Settled agriculture, which had been impossible in the wild swings of previous times, emerged. Regular floods following seasonal snow-melt in the mountain sources of the great rivers allowed irrigation to multiply crop yields on fertile soils. The breakthrough began some 7,000 years ago in the 'cradle of civilization', the river valleys that drain into the Persian Gulf.[7] The new conditions permitted the development of the wheel, writing, mathematics, legal codes, centralized government and social strata. It is for this reason that we can speak of the 'halcyon millennia of the Holocene'.[8]

Now, the Holocene has come to an end. Humans have flourished so successfully in the sympathetic environment of the last 10,000 years that they have shifted Earth's geological arc. The impact of burning fossil fuels on the Earth's atmosphere has been so far-reaching that it is the principal factor, along with population growth (up from 800 million in 1750 to 7 billion today), that has persuaded Earth system scientists to declare that the Earth has entered a new geological epoch known as the Anthropocene.[9] The post-Holocene epoch is defined by the fact that the 'human imprint on the global environment has now become so large and active that it rivals some of the great forces of Nature in its impact on the functioning of the Earth system'.[10] It was Paul Crutzen who in 2000, with ecologist Eugene Stoermer, first announced its arrival. In a seminal intervention, they suggested that the Anthropocene, the 'Age of Humans', may be said to have begun in 1784 with the commercialization of James Watt's steam engine.[11] As we will see, in a scientific debate with profound implications the starting date of the Anthropocene is in dispute. Whatever the case, the Anthropocene entered a turbocharged phase in the 1950s when greenhouse gas emissions accelerated sharply. A plethora of indexes of resource consumption, waste output and environment degradation show a sudden and sustained acceleration from that decade.

In one of the strongest interpretations of the new epoch, Erle Ellis writes that in the Anthropocene

the terrestrial biosphere is now predominantly anthropogenic, fundamentally distinct from the wild biosphere of the Holocene and before . . . [N]ature is now human nature; there is no more wild nature to be found, just ecosystems in different states of human interaction, differing in wildness and humanness . . .[12]

By any but the purest definition of wilderness, this claim is excessive. Studies vary, but the Wildlife Conservation Society estimates that 26 per cent of the Earth's land surface may be classed as 'last of the wild'.[13] And in what only appears to be a contradictory sign, there remain perhaps a hundred uncontacted tribes in the Amazon and West Papua. As we will see, Ellis represents a new Promethean ecological politics that welcomes the disappearance of wilderness because it seems to justify human domination.

Nevertheless, the trend is unmistakable and Ellis is on surer ground when he adds: 'By the latter half of the twentieth century, the terrestrial biosphere made the transition from being shaped primarily by natural biophysical processes to an anthropogenic biosphere in the Anthropocene, shaped primarily by human systems.'[14] Ellis is preoccupied with the landscape and biosphere; but the principal motivation for announcing the arrival of the Anthropocene comes from human transformation of the atmosphere.

The climate under the enhanced greenhouse effect is expected to be much less clement than the one to which we have grown accustomed. This is apparent from the intense focus of climate researchers on feedback effects, tipping points, extreme weather events, abrupt climate change and climate emergencies, all driven by rates of atmospheric carbon dioxide and temperature increase unprecedented in the palaeoclimate record. Greater variability, in the form of heatwaves, floods, cyclones and so on, is to be expected when more energy, in the form of heat, is stored in the atmosphere. In a landmark intervention in 2009, 27 experts wrote in *Nature*:

Many subsystems of Earth react in a nonlinear, often abrupt, way, and are particularly sensitive around threshold levels of certain key variables. If these thresholds are crossed, then important subsystems, such as a monsoon system, could shift into a new state, often with deleterious or potentially even disastrous consequences for humans.[15]

Contrary to the comforting conception of robust Nature, these scientists believe the upheaval of the Anthropocene 'could see human activities push the Earth system outside the stable environmental state of the Holocene'. The focus on past resilience may 'lull us into a false sense of security because incremental change can lead to the unexpected crossing of thresholds that drive the Earth System'.[16]

The Anthropocene is now the subject of extensive scientific investigation, but almost no thought has yet been given to its larger meaning. On the one hand, modern humans have acquired an extensive array of know-how, a book of blueprints that could help protect us from a climate shift. After all, technological civilization has been driven in large measure by the desire to isolate ourselves from the vagaries of the weather. Yet our ability to survive is hostage to a vast network of static infrastructure – water supplies, power grids, industrial estates, road systems, central business districts, suburbs, agricultural systems – all designed to function smoothly in the climatic conditions of the Holocene, and whose fragility in the face of extreme weather events is all too apparent to anyone with a television set. For all of our fantastic achievements, in the face of a rapid climate swing we may discover we have feet of clay.

The fact that the infrastructure for 7 billion people to live as they do today has taken several hundred years to build (a few thousand if we include agriculture), and has been possible because of the relatively stable and sympathetic climate that marked the Holocene, is an awkward truth for those who argue that the Earth's climate has always changed and humans have survived so there is no need to worry about more change. If survival of the species is the only goal then there is perhaps no reason for alarm. But for those who value civilized society and who are not willing to turn their faces away from the poorest and most vulnerable people of the world, the reasons to fret are numberless.

Suppressing ice ages

That the persistence of excess carbon dioxide in the atmosphere is the most profound and least understood fact about human-induced climate change has already been stressed. The climatic effects of burning fossil fuels will last longer than Stonehenge, longer even than nuclear waste.[17] Thus the future of the Earth over the coming millennia is already inscribed in the atmosphere, with more disturbance locked in by fixed patterns of thought that will see us release more greenhouse gases in the next decades, etching the future more deeply into the sky. The long-term effects are understood well enough for palaeoclimatologists to make projections for the evolution of Earth's climate over the next tens of thousands of years. This astonishing emerging story, one that throws everything we understand about human beings into a new light, is told in two recent books by palaeoclimatologists David Archer and Curt Stager.[18] The story goes like this.

Over the last 34 million years, since the Antarctic ice sheet formed, the Earth's climate has fluctuated around a pattern of long glaciations punctuated by warm periods. Compared to the ice ages the warm periods have been brief, around one-tenth of the duration of an ice age, short enough for them to be dubbed 'interglacials'. After a century or so of close study, the mechanisms behind these fluctuations are well, if not completely, understood. The great glacial succession responds to three primary cycles, each of which influences the amount of solar energy or insolation reaching the planet.[19] In calculating the effects, changes in the intensity of sunlight hitting the top of the Earth (north of latitude 65°) in the summer months are especially significant because the northern zone has the most influence on the amount of ice on the planet. As Archer puts it: 'The northern hemisphere summer is the solar-forcing sweet spot that drops the entire planet into an ice age', or bounces it back out, which helps to explain why climate scientists watch the Arctic so nervously.[20] Ice formation at the top of the Earth is much more dynamic than it is at the bottom.

The first of the three cycles is the Earth's wobble around its axis of rotation. Occurring roughly every 23,000 years, this *precession* cycle changes the amount of solar radiation reaching the northern hemisphere (with an equal and opposite amount reaching the southern hemisphere but with less impact). The second cycle is the Earth's tilt, or *obliquity*, which changes how far the Earth's axis leans away from its orbit. A greater lean means that seasons are more distinct. This tilt from side to side occurs with regularity over a period of 41,000 years.

The third is the *eccentricity* cycle. The Earth's annual orbit around the Sun varies from nearly circular to somewhat elliptical. When it is more circular the amount of sunlight reaching the Earth is more constant through the year. This cycle occurs over a period of approximately 400,000 years, with a shorter cycle of around 100,000 years within it.

The three cycles interact and together are largely responsible for the arrival and cessation of ice ages. Using knowledge of these cycles climatologists are able to predict that Earth is due for its next ice age in about 50,000 years' time – although our confidence in all such predictions is eroded by the instability arising from human disturbance to the Earth system. Such an ice age would normally be associated with a concentration of carbon dioxide in the atmosphere of around 250 ppm. However, fossil fuel emissions over the last century or so have raised the concentration from 280 ppm before the industrial revolution to 395 ppm today, and over the next decades will raise it to at least 550 ppm and perhaps as high as 1,000 ppm. This greenhouse gas blanket will cause the planet to warm for a very long time. In fact, palaeoclimatologists expect that the persistence of carbon dioxide in the atmosphere means that releasing 1,000 billion tonnes of extra carbon into the atmosphere (we are halfway there already) *will suppress the next ice age*. If emissions rise to the higher level of 5,000 billion tonnes of carbon (by burning all known fossil fuel reserves) then a couple of centuries of industrial activity could stem the subsequent ice age as well, expected in about 130,000 years, and indeed all glaciations for the next half a million years.[21]

Nothing humans have ever done approaches the momentousness of this conclusion. Our activities have so changed the climatic future that we have prevented one and possibly several ice ages. I think it will take decades for us to understand the enormous implications and meaning of this fact. The Earth will take tens or hundreds of thousands of years to reach a new equilibrium following the pulse of carbon emissions sent into the atmosphere by humans mostly over the century from 1950 to 2050. Only then could the era of human-induced global warming come to an end.

Converging histories

Climate change is destabilizing the modern conception of the Earth as a complex system whose secrets can be known and whose course can be foreseen. Science itself is pointing towards the inherent inscrutability of the natural world. We have seen how global warming is affecting the biosphere, the hydrosphere and the cryosphere. Scientists are now beginning to grasp the way in which human-induced climate change can affect the lithosphere (the outer crust of the Earth) and the geosphere (the deeper structures of the planet). It is now emerging that by shifting the distribution of ice and water over the surface of the Earth, human-induced global warming is likely to provoke geological and geomorphological responses, including seismic, volcanic and landslide activity.[22] Changes in the seasonal snow-load, for example, affect seismic activity in Japan by changing the compression on active faults. According to a recent scientific review of the field, in Iceland and Alaska 'melting of ice in volcanic and tectonically active terrains may herald a rise in the frequency of volcanic activity and earthquakes'.[23] Volcanoes can cool the planet, but a warming planet can also trigger volcanoes. When glaciers melt, the earth 'rebounds'. With a decline in ice load of 1 kilometre, the Earth's crust may rise by hundreds of metres. Moreover, although effects on the climate and biosphere are far more important impacts, the melting of polar ice due to global warming and the consequent redistribution of the weight of the Earth's water can be expected to alter slightly the Earth's rotation speed and its orientation in the solar system.[24] In the words of Bill McGuire, the geophysicist who wrote the book on the geological implications of climate change, our knowledge reveals:

> the all-encompassing and all-pervasive nature of rapid and severe climate change. So complex and entangled is the Earth System that, looking to the future, nothing can be regarded as immune to the influence of anthropogenic warming . . . we are already seeing the first signs of the geosphere responding to changes wrought by rising temperatures.[25]

The point of all this is that the effects of human-induced warming go far beyond changes in the weather; *everything* is now in play, and not only scientifically. So let us now make a leap from the land of science to the grounds of the humanities, because the astounding new facts uncovered by Earth system science force us to rethink our understanding of history.

The idea that humanity makes its own history and does so against the backdrop of the Earth's slow unconscious evolution is deeply implicated in modernity. We are accustomed to thinking of humans, having emerged from the primordial darkness, as independent entities living and acting on a separate physical world, a world we plough up, mine, build on and move over but which nevertheless has an independent existence and destiny. This understanding of the autonomy of humans from nature runs deep in modern thinking; we believe we are rational creatures, arisen from nature, but independent of its great unfolding processes. To be sure, human exceptionalism has an ancient lineage; but before the Enlightenment the special place humans occupied was contained in a unified cosmos and it was God who blessed humans with uniqueness. The mark of the modern world was that humans designated themselves the unique species. The theory of evolution was implicated in this view of human distinctiveness, but the modern idea of human exceptionalism emerged only with the science of *geology*. It was geology that gave the Earth a history and it was only after nature acquired its own history that humans could acquire theirs.

Before the Enlightenment the historical process was understood to be identical with the unfolding of God's purpose. In 1605 Francis Bacon famously argued the doctrine of the 'two books', the book of Nature and the book of Scripture, successfully splitting off the natural world for a new and different mode of understanding, *eruditio* as distinct from *divinatio*, which would later that century evolve into the scientific method.[26] Bacon argued that we should apply our understanding of things in nature in order to interpret the words of the Scripture, rather than drawing on Scripture to interpret nature – just as Galileo had argued that we should use physics for scriptural exegesis, to reveal 'the true senses of the Bible'.[27]

Since in Genesis the origins of the Earth and the origins of the cosmos were seen to be concurrent, a vital stage in the emergence of a distinct human history was the separation of the history of the Earth from the history of the cosmos, a task initiated by Giordano Bruno and taken up by René Descartes, who suggested a scheme in which the Earth had its own origin. Moves like this allowed the development of cosmogenic stories that combined geological knowledge with Scripture in what became known in the eighteenth century as 'theories of the earth'.

The emerging discipline of geology – marked by the new scientific emotions of sobriety and detachment – adopted Bacon's compromise so that the history of the Earth could be separated out from the biblical narrative. Freed from fealty to Genesis, natural scientists could pursue their own method of inquiry, and use their results, if they were so inclined, to explicate and fill gaps in the scriptural story.[28]

One consequence was that human history, which had been enfolded with natural history into the comprehensive account of the cosmos found in Genesis, began to be more clearly distinguished too. The narrative of Genesis, wrote Martin Rudwick, made it 'plausible to regard human history as virtually coextensive with earth history; without mankind the earth and the cosmos would have seemed to lack meaning and purpose', so before the emergence of modern geology the 'history of the earth was seen only as a stage for the drama of human history, the drama of the creation, fall, and redemption of a unique set of rational beings'.[29] In the seventeenth century Sir Thomas Browne could unselfconsciously remark: 'Time we may comprehend, 'tis but five days elder than ourselves.'[30]

In Britain, Bacon's 'two books' compromise lasted into the nineteenth century, at which stage scriptural loyalists found that, as new and persuasive chapters were added to the book of science, the authority of the Bible could be bolstered only by more subtle exegesis or a new revelation, which proved elusive. Reconciling biblical accounts with newer discoveries in astronomy and the fossil record became increasingly awkward. The fossil evidence could no longer be accounted for by the Flood, and accumulating proof for the antiquity of man suggested a human history that pre-dated Adam.[31] The discovery that beneath the 'secondary' rock strata that contained fossils there were 'primary' rock strata devoid of fossils, including human ones, implied an earth history prior to humans, and one of very long duration. Earth had a history before the arrival of humans; indeed, before the arrival of life.

It was these discoveries that eventually led Charles Lyell in 1830 to declare that his aim was to 'free the science from Moses'.[32] It was still possible for those so inclined to argue that all stages of the Earth, including the pre-human ones, were the result of divine providence, and thus part of God's design. Nevertheless, such a view seemed to rely more on faith than evidence, and evidence now mattered more. Darwin's theory on the origin of species told a story of life developing against the backdrop of an Earth history following its own course. Darwin's ideas were built on geology and it was the palaeontologists, combining biology with geology, who gave the Earth a history that could encompass evolution. The prising of Earth history away from a larger religious cosmology allowed the flourishing of a new human history too, driven not by divine purpose but by mundane forces like states, empires, techno-logical change, class conflict and economic growth.[33]

In this sketch of geology's emergence we can see that Earth history and human history have taken separate courses for not much more than two centuries.[34] The bifurcation was an essential moment in the evolution of the modern subject, the autonomous agent acting on the external world. The autonomous subject, taken collectively, must have an autonomous history. Moreover, human history acquired the quality of *progress*, while that of the Earth lost its telos. Darwin was to supply it with an unconscious evolutionary dynamic. Even so, the evolution of humans has nothing to do with history but lies in the province of 'deep history'. In 1964 E. H. Carr expressed the universal modern view when he wrote: 'History begins when men begin to think of the passage of time in terms not of natural processes – the cycle of the seasons, the human lifespan – but of a series of specific events in which men are consciously involved and which they can consciously influence.'[35] It is true that in more recent times environmental historians have emphasized how nature and natural events have always shaped human affairs, so that human history can never float entirely free of the Earth's constraints. Nevertheless, those ideas have not yet dented the essential modernist belief, expressed in the nineteenth century by Jacob Burckhardt, that history is 'the break with nature caused by the awakening of consciousness'.[36]

In an important observation, historian Dipesh Chakrabarty has pointed out that the distinction we have drawn between *natural* history – slow processes that occur on a scale of millions of years – and *human* history – a series of events that occur on the scale of years, decades and centuries – has now collapsed.[37] With the Anthropocene, humans have become a geological force so that the two kinds of history have converged and it is no longer true that 'all history properly so called is the history of human affairs'.[38] Our future has become entangled with that of the Earth's geological evolution. Anthropogenic climate change affects not just the atmosphere but the chemical composition of the oceans (acidification), the biosphere (species extinctions and shifting habitats), the cryosphere (melting ice masses) and the lithosphere itself.

The force of this is redoubled when we remember that the long-lasting effects of increased atmospheric carbon dioxide mean that human activity is likely to suppress the next ice age, so that the two histories are inseparable for at least that time. It turns out that the 'clever animal' who managed to separate itself from nature so completely that it could acquire its own history has so transformed the Earth that it now peers nervously into a future of unaccustomed instability and danger. Contrary to the modernist faith, it can no longer be maintained that humans make their own history, for the stage on which we make it has now entered into the play as a dynamic and largely uncontrollable force.

A good Anthropocene?

One response to the arrival of the Anthropocene, the usual one, is to argue that it has arisen because of a regrettable failure of foresight. We have not given enough thought to the side effects of our technological progress, so to save the situation we need better scientific understanding and technological know-how. On this view, the response to the climate crisis and the broader dangers presented by the Anthropocene lies in raising to a higher level the characteristic of humans that makes us distinctive as a species, our reasoning capacity. The disruption of the Anthropocene demands that we redouble our belief in the perfectibility of humankind. Yet how can we think our way out of the problem when the problem is the way we think?[39] There is something increasingly desperate about placing more faith in technological cleverness when it is the unrelenting desire to command the natural world that has brought us to this point. Unless we understand why a certain kind of rationality seems to have failed, appeals to more reason are quixotic. After all, the separation of natural and human history and the dominance of a certain form of calculative rationality were each products of the same Enlightenment process.

The type of thinking embedded in the framework of systems analysis, risk assessment and cost–benefit calculation can be called 'technological thinking'. Technological thinking understands the world as a collection of more or less useable resources. According to this view, technology transforms potentially useful things into useful things without asking about the origins of the world as an assortment of potentially useful things. As such, modern technology reveals something essential to the nature of modern humans – the determination to shape the world around us to suit desires that seem to have no limit.

Plans to engineer the Earth through the deployment of contrivances to manipulate the atmosphere represent the fulfilment of three and a half centuries of objectification of nature. The Earth as a whole is now represented no longer simply as a collection of objects but as an object in itself, one open to regulation through the 'management' of the amount of solar radiation reaching the planet. So we begin to consider suitable political institutions for regulating the amount of light reaching the planet, a task that since the formation of the Earth 4.5 billion years ago has been left to the Sun mediated by the Earth's atmosphere. It seems we are no longer happy with the arrangement and want to assume control ourselves. Earth-as-object also underlies the idea that we can adjust the volume of greenhouse gases in the atmosphere to a level calculated to be 'optimal'. In this view, why wouldn't we make plans to prevent the next ice age, due in 50,000 years' time? Curt Stager himself reflects on how science can inform 'long-range planetary management' which might include leaving some coal in the ground so future generations can burn it, 'in a responsibly controlled manner', to keep an ice age at bay.[40]

In their 1997 paper 'Global warming and ice ages: Prospects for a physics-based modulation of global change', Edward Teller, Lowell Wood and Roderick Hyde developed a proposal to eliminate what they call 'climate failures'.[41] The Earth's climate system 'fails' when it no longer provides the requisite functionality, its requisite function being to satisfy the needs of humans. It may be necessary, they argue, to suppress ice ages by pumping more greenhouse gases into the atmosphere, or by deploying a cloud of sunlight-deflecting particles to a position slightly offset from the Earth–Sun axis to direct additional sunlight on to the Earth. After all, while greenhouse warming is only 'a possibility', ice age cooling 'is a practical certainty'. Perhaps it is mere idle speculation or scientific hubris to consider plans for overruling the next ice age when the arrival of that event is 25 times more distant from today than the birth of Christ. When asked to respond to the coming of the next ice age, Prometheans think of technological interventions; Soterians are inclined to reply: 'Ask me in 49,000 years time.' Yet the need to consider how to deal with the next ice age is being used to justify regulating the climate now; we will need the technology at some point, Teller and Wood reason, so let's go. For the true Prometheans it is not enough to regulate today's climate; the goal is to take control of geological history itself. To the Earth they repeat the words of the creature to Dr Frankenstein: 'You are my creator, but I am your master.'

For some, the lesson to be learnt from the arrival of the Age of Humans is not the need for greater humility but its opposite, an invitation to assume total control. Throwing off all small-minded fears, their slogan is 'Welcome to the Anthropocene'. Here we come to a debate in the scientific literature with the most far-reaching implications. The debate is over when the Anthropocene began.

Contrary to the claim that the shift occurred with the industrial revolution, palaeoclimatologist William Ruddiman argues that the Anthropocene began some 8,000 years ago with the onset of forest clearing and farming, which led to enhanced levels of methane and carbon dioxide in the atmosphere.[42] Paul Crutzen and Will Steffen defend the claim that it properly began in the late eighteenth century with evidence showing that human impact on the world as a whole was not discernible 7,000 or 8,000 years ago, and certainly was not large enough to upset the stability of the Holocene. Looking at the record it is indisputable that a step change occurred in the late eighteenth century, the beginning of the Industrial Revolution, the event that 'unbound Prometheus' and sparked the modern urge to mastery over nature. Beyond that step, the charts also show a startling leap after World War Two. 'The mid-twentieth century was a pivotal point of change in the relationship between humans and their life support systems,' write Crutzen and Steffen. 'The period of the Anthropocene since 1950 stands out as the one in which human activities rapidly changed from merely *influencing* the global environment in *some* ways to *dominating* it in *many* ways.'[43]

The dispute is not merely academic. One implication of Ruddiman's 'early Anthropocene' argument is that if humans have been a planetary force since civilization emerged, then there is nothing fundamentally new about the last couple of centuries of industrialism. It is in the nature of civilized humans to transform the Earth, including by the use of geoengineering, and what is in the nature of the species cannot be resisted. By focusing attention on 'humankind' in general rather than forms of social organization that emerged more recently, the Anthropocene becomes in some sense natural. It is not the product of industrial rapaciousness, an unregulated market, human alienation from nature or excessive faith in technological power; it is merely the result of humans doing what humans are meant to do, that is, use the powers Prometheus gave us to better our lot. If humankind is in this sense inseparable from Nature then there is nothing inherently preferable about the natural state or the Holocene climate. Thus in one reading of the ethics of geoengineering, 'there is no prima facie justification for attempting to preserve the current climate, if some other climate might be better for humans and animals'.[44] Depending on the assessment of human well-being (and perhaps other sentient beings), there may be a 'better' temperature or climate as a whole. It is therefore justified for humans to 'set the global thermostat' wherever we please.

Erle Ellis, an ally of Ruddiman in the debate, defends the 'good Anthropocene'. He is confident 'human systems' can adapt and indeed prosper in a hotter world because history proves our flexibility. There are no planetary boundaries that limit continued growth in human population and economic advance. The Anthropocene is barely distinguishable from the Holocene; the only barrier to a golden future is human self-doubt.

> A good ... Anthropocene is within our grasp. Creating the future will mean going beyond fears of transgressing natural limits and nostalgic hopes of returning to some pastoral or pristine era. Most of all, we must not see the Anthropocene as a crisis, but as the beginning of a new geological epoch ripe with human-directed opportunity.[45]

Humans look for ways to welcome that which seems inevitable. Ellis is in the vanguard of a movement to re-imagine the Earth, a cheerful vision articulated by science writer Emma Morris in her book *Rambunctious Garden*. Instead of lamenting the loss of wilderness, we remake it: 'Rambunctious gardening is pro-active and optimistic; it creates more and more nature as it goes, rather than just building walls around the nature we have left.'[46]

For Ellis, Morris and those of like mind, humanity's transition to a higher level of planetary significance is 'an amazing opportunity'. Ellis expects that 'we will be proud of the planet we create in the Anthropocene'.[47] In his embrace of the benevolent Anthropocene, and in a foretaste of a conservative reframing, Ellis is joined by Ronald Bailey from the libertarian *Reason* magazine, who believes we can only become better at being the 'guardian gods of Earth'.[48] The early Anthropocene hypothesis is interpreted as exonerating modern humans from blame for environmental decline. The new epoch is read as the 'manifest destiny' of humanity, a reading that finds a more sympathetic ear in the United States than in Europe. Even so, this kind of American Promethean dreaming meets stiff resistance in its homeland; the *New York Times* opinion pieces in which Ellis and Bailey expressed their views were met with a barrage of Soterian objections: 'nothing to be proud of'; 'the Anthropocene era may be extremely short-lived'; 'we have not much control over what mother nature has in store for us next'; 'Ozymandias'; and so on.

Perhaps the defenders of the 'good Anthropocene' intuitively understand that if the beginning of the new epoch is located at the end of the eighteenth century, with a step-change in the 1950s, we must ask what was distinctive about those times. The answer of course is the inception of industrial capitalism and then the turbo-charged era of expansion that followed World War Two, a surge only intensified by the outbreak of consumerism that washed over the rich world in the 1990s and 2000s. If industrial rapacity and 'affluenza' are the source of the problem then perhaps the system's inclination to excess can be curbed, or even reversed, a conclusion from which conservatives instinctively recoil.

If by looking back 8,000 years the defenders of the good Anthropocene can allay their anxiety with a redoubled confidence in human creativity, a deeper form of stress relief is available to those who look back even further. The more one studies the deep history of the Earth the more a sense creeps over the mind that everything we have built is ephemeral and insignificant and will inevitably be overwhelmed by the great geological processes that lift mountain ranges, split continents and cause sea levels to plunge. Humanity becomes a species that arose because conditions arrived that suited it, a species that sometimes flourished and, as conditions turned hostile, sometimes struggled to survive. In this mode of thinking, politics, suffering, hopes and dreams – in short, the lives of actual mortals – lose their meaning. It lends itself to a kind of *palaeofatalism*, an existential complacency that settles over those who spend too long immersed in geological timescales. It helps explain why so many geologists are indifferent or hostile to the warnings issued by climate scientists.[49] Curt Stager has fallen victim to it: 'A deeply historical perspective can make modern greenhouse heating seem no more outlandish that the natural PETM and Eemian warm periods of the distant past . . .'[50] Of course, when the Earth suddenly went into hothouse conditions in the Palaeocene–Eocene Thermal Maximum (PETM) 55 million years ago there were not 7 billion mortals living on it. 'Why fear change when we live in such an inconstant world?' Stager asks. 'I sometimes wonder if it's not global warming that worries us so much as change of any kind'[51] – an indifference to suffering perhaps reserved for those best placed to survive in a world of climate change.

Growth fetishism

In his 1784 essay 'What is enlightenment?', Immanuel Kant captured the anti-clerical mood sweeping Europe with the motto

'Sapere aude!' – Dare to know![52] The ideas of the Enlightenment spread from Europe to America through the writings of Benjamin Franklin and Thomas Jefferson, among other republicans, and the United States subsequently became the powerhouse of modern scientific achievement. Yet it now seems that science could remain pre-eminent only as long as its knowledge served a deeper purpose: unceasing expansion. Today the motto of those US conservatives who have turned against science might be 'Fear to know'! It is a fear that finds its most primitive expression in declarations such as that of US Senator James Inhofe: 'God's still up there. The arrogance of people to think that we, human beings, would be able to change what He is doing in the climate is to me outrageous.'[53]

Beyond understanding them as mere human weakness or distorted expression of political objectives, I suggest that the kinds of denial and evasion that have led us to the point of contemplating geoengineering are means of attempting to resolve the contradiction deep within the modern understanding of the world itself. The contradiction arises because science has thrown up some facts that challenge the foundation of the modern understanding of the world, that is, the conception of humans as self-determining agents able to control the future by exercising power over nature. Climate engineering and the 'good Anthropocene' hypothesis can be understood as an attempt to resolve this contradiction.

The pressure has been building for some decades now. The disproportionate rage that greeted the publication of *The Limits to Growth* in 1972 begins to become explicable when we understand just how deeply rooted in the modern Western soul is the expectation of endless expansion. Growth has become fetishized, that is, invested with magical powers. Growth is the modern world's most powerful emblem – the symbol of virility, of the future, of life itself. Continuing growth provides the ballast for our dreams and our hopes. The distant and abstract language of systems analysis used by the MIT authors of *The Limits to Growth*, language that had been appropriated by techno-expansionism, only added to the sense of betrayal many felt at its stunning conclusion – that humanity must 'begin a controlled, orderly transition from growth to global equilibrium'.[54] The truth is that restricting global greenhouse gas emissions to something approaching a safe level is an impossible burden for technology to carry.[55]

'Hell is the impossibility of expanding,' observed the philosopher Peter Sloterdijk.[56] To question the possibility of, let alone the need for, endless expansion is to commit the modern sin. Those who seem to challenge this most holy of truths must be cast out. This is the only explanation I can think of for the extraordinary hostility directed by some at environmentalists, who are now regularly accused of being *responsible* for environmental decline.[57] They have become the scapegoats for our sins, loaded up with our own guilt and sent off to the wilderness. Indeed, those environmentalists who have taken themselves off to the wilderness – to live more simply and grow organic foods – have cast *themselves* out and so become easy targets for ridicule and vilification.

The disclosure of the Anthropocene – and the paramount fact of the persistence of carbon dioxide in the atmosphere – means that the grand narrative of the Enlightenment – that of unending progress achieved through the application of human ingenuity applied to an inert external world – can no longer hold up. As the basis for our deepest social structures, as well as our individual understandings of our own futures, the destabilization of the narrative calls everything into question. The direction in which we thought we were going has now been denied to us. The historical force of this should not be missed, for it means that the utopian promise of all political and religious ideologies, both materialist and metaphysical, vanishes.

So we have reached the point in history where we must face up to the tragic consequences of 'the American way of life', a way of life also lived in other affluent countries, albeit typically with less intensity and ideological conviction. The same qualities that made the United States a great nation – relentless optimism, commitment to know-how, determination to expand – have become the enemies of its preservation and, collaterally, the preservation of the rest of humanity. A nation that has expansion running in its blood can barely conceive of contraction, and so the question we will soon be forced to ask is how much of the rest of the world will be *sacrificed* to prolong the dream of affluence?

We have seen that it is not true that Prometheans must favour climate engineering and Soterians must oppose it. Nor is it true that Soterians are against technology. It is not so simple. Yet among those who believe we should make preparations to engineer the climate there is a sharp division between Prometheans and Soterians. The former are inclined to see it as a way of defending the established order so that expansion can continue uninterrupted. The latter see it as a regrettable measure to protect those deeper values now threatened by the consequences of endless expansion – viable societies, vulnerable communities, ecological values and life itself. So there are two distinct questions we must face with climate engineering: should we undertake it and, if we do, how do we use it? Do we use it to 'save' growth and so jeopardize those deeper values because it can only postpone the reckoning; or do we use it to protect those values while we wean ourselves from endless growth? I hope I have said enough about climate science to convince the reader that using geoengineering to defend continued expansion cannot work in the long term. The only justification for deploying geoengineering is to make it easier politically to transform our economies and societies so that we live in a way that does not disrupt Earth's natural cycles and the processes that have allowed life to flourish. So if it comes to pass that a World Climate Regulation Agency is created, then prudence demands that it be staffed by Soterians, and that every advertisement for positions at the agency be marked with the words 'Prometheans need not apply'.

Yet the prospects of this occurring are slim; the agents of Prometheus are colonizing climate engineering. We can see in embryo a lobby that unites fossil fuel corporations opposed to carbon reduction policies with investors in geoengineering technologies. The two could soon overlap. We see powerful forces of denial in the United States and elsewhere shift to supporting geoengineering, and conservative political leaders beginning to see the electoral advantages. The high priests of the Prometheus cult, the free market economists, are naturally drawn to it. The strategic significance of climate engineering is likely to lead to its progressive militarization. Technological thinking structures our consciousness in a thousand subtle ways that make climate engineering attractive, indeed, almost inevitable.

Prometheans rule. Over three centuries of advance, displaced workers, romantic poets, dismayed clerics and far-seeing ecologists put up resistance; all sooner or later were crushed. Who can hold back such a force? Yet history proves that the invincible can be thwarted and the mighty brought to heel in unexpected ways. As the Chinese proverb has it: when taken to their extreme, things revert to their opposite. Only history can answer whether the time has come; but if the meek are ever to inherit the Earth then they had better be quick.

Hubris, profligacy, weakness of will, power-hunger, fear, wilful blindness, false hope and the capriciousness of Nature – titanic forces are at work in the world. Ovid's story of Phaëton echoes down the centuries. To prove his love, Helios promised his son Phaëton anything he desired. When he asked to drive his father's chariot, the Sun, for a day, Helios balked at the thought of this callow youth holding the reins of such power. But he had made a promise. When Phaëton took charge of the chariot he lost control of the horses, whose wild path across the heavens first froze the Earth then scorched it.

> Th' astonisht youth, where-e'er his eyes cou'd turn,
> Beheld the universe around him burn.

To bring the chaos to an end, Zeus aimed a thunderbolt at Phaëton, who plunged to earth.

> And o'er the tomb an epitaph devise:
> 'Here he, who drove the sun's bright chariot, lies;
> His father's fiery steeds he cou'd not guide,
> But in the glorious enterprize he dy'd.'

In the next decades we will discover whether attempting to engineer the climate is glorious enterprise or ruinous folly, whether Prometheus will crow or Soteria will weep.

Notes

Chapter 1 Why Geoengineering?

1 Rafael Romo, 'Whitening mountains: A new effort to save Peruvian Andes glaciers', *CNN U.S.*, 28 Nov. 2011, at http://articles.cnn.com/2011-11-28/americas/world_americas_peru-mountain-whitening_1_mountains-alpaca-ice?_s=PM:AMERICAS (accessed Jan. 2012).

2 'Keep Earth cool with moon dust', *New Scientist*, 9 Feb. 2007, at http://www.newscientist.com/article/dn11151-keep-earth-cool-with-moon-dust.html (accessed Jan. 2012).

3 A. B. Kahle and D. Deirmendjian, 'The black cloud experiment', Report R-1263-ARPA, RAND Corporation, Santa Monica, CA.

4 Chalmers Johnson, 'A litany of horrors: America's university of imperialism', 29 Apr. 2008, at http://www.tomdispatch.com/post/174925/chalmers_johnson_teaching_imperialism_101 (accessed Jan. 2012). Johnson was once a consultant to RAND.

5 P. C. Jain, 'Earth-Sun system energetics and global warming', *Climatic Change*, 24 (1993), pp. 271–2.

6 Ibid., p. 272.

7 See https://groups.google.com/group/geoengineering/browse_thread/thread/660846de67b3a26c (accessed Jan. 2012). Geoengineering advocate Andrew Lockley describes asteroid nudging as 'a fun idea'.

8 D. Price, 'Is man becoming obsolete?', *Public Health Reports*, 74 (Aug. 1959), p. 693.

9 Ibid., p. 694.

10 Foreword to G. Morgan and K. Ricke, 'Cooling the Earth through solar radiation management: The need for research and an approach to governance', opinion piece for International Risk Governance Council, 2010.

11 International Energy Agency, *World Energy Outlook 2011* (Paris: IEA, 2011), p. 2.

12 The best attempt is by Mark Lynas in *Six Degrees: Our Future on a Hotter Planet* (London: Harper Perennial, 2007).

13 Jane Risen and Clayton Critcher, 'Visceral fit: While in a visceral state, associated states of the world seem more likely', *Journal of Personality and Social Psychology*, 100:5 (2011).

14 Erica Dawson and Kenneth Savitsky, '"Don't tell me, I don't want to know": Understanding people's reluctance to obtain medical diagnostic information', *Journal of Applied Social Psychology*, 36:3 (2006).

15 Ross Garnaut, *The Garnaut Climate Change Review* (Melbourne: Cambridge University Press, 2008), table 3.1, p. 56.

16 René Girard, *Battling to the End* (East Lansing: Michigan State University Press, 2010), p. xiii.

17 Bill McKibben, 'Global warming's terrifying new math', *Rolling Stone*, 2 Aug. 2012.

18 Guy Pearse, 'Mine coal, sell coal, repeat until rich', Dec. 2011, at http://www.guypearse.com/docs/guypearse.com/Woodford%20Dec%20FINAL%20%20 2011.pdf (accessed Jan. 2012).

19 Personal communication. One or two papers casting doubt on one or more of the main propositions of climate science have appeared in peer-reviewed journals. Unfortunately, their analyses or data have soon been shown to be faulty. In one case, the editor resigned for allowing through a paper that did not meet the analytical standards.

20 J. Screen and I. Simmonds, 'The central role of diminishing sea ice in recent Arctic temperature amplification', *Nature*, 464 (29 Apr. 2010), pp. 1334–5.

21 This cannot be true indefinitely as warming must have an upper bound (perhaps 9–10°C warmer than at present), so in the very long term (tens of thousands of years) powerful negative feedback effects must come to dominate, restoring the Earth to a stable, if very different, climate.

22 W. Broecker, 'Ice cores: Cooling the tropics', *Nature*, 376 (20 July 1995), pp. 212–13.

23 Committee on Abrupt Climate Change, National Research Council, *Abrupt Climate Change: Inevitable Surprises* (Washington, DC: National Academy Press, 2002).

24 Robert B. Gagosian, 'Abrupt climate change: Should we be worried?', prepared for a panel on abrupt climate change at the World Economic Forum, Davos, Switzerland, 27 Jan. 2003, at http://www.whoi.edu/page.do?pid=83339& tid=3622&cid=9986 (accessed Sept. 2012).

25 Johan Rockström et al., 'A safe operating space for humanity', *Nature*, 461 (24 Sept. 2009), pp. 472–5.

26 Paul Crutzen, 'Albedo enhancement by stratospheric sulfur injections: A contribution to resolve a policy dilemma?', *Climatic Change*, 77 (2006), pp. 211–20.

27 Mark Lawrence, 'The geoengineering dilemma: To speak or not to speak?', *Climatic Change*, 77 (2006), pp. 245–8.

28 'Climate scientists: It's time for "Plan B"', *Independent*, 2 Jan. 2009.

29 Chris Mooney, 'Copenhagen: Geoengineering's big break', *Mother Jones*, 14 Dec. 2009.

30 John Vidal, 'Geo-engineering: Green versus greed in the race to cool the planet', *Observer*, 10 July 2011.

31 David Victor, M. Granger Morgan, Jay Apt, John Steinbruner and Katharine Ricke, 'The geoengineering option: A last resort against global warming?', *Foreign Affairs*, 88:2 (Mar./Apr. 2009).

32 James Hansen, 'Heroes of the environment', *Time*, 17 Oct. 2007.

33 Although a case could be made for Metis, the goddess of prudence, wisdom and cunning, or Aidos, the goddess of shame, modesty and humility.

Chapter 2 Sucking Carbon

1 Graeme Pearman, derived from Global Carbon Project 2010, at http://www.globalcarbonproject.org/carbonbudget/index.htm (accessed Apr. 2012).

2 David Archer, *The Long Thaw* (Princeton: Princeton University Press, 2009), p. 109.

3 Ibid., p. 123.

4 Ibid., table 2, p. 117.

5 There are around 55 million GtC stored in the lithosphere. It is periodically vented into the atmosphere by volcanic eruptions. See Roger Gifford, 'Global photosynthesis in relation to our food and energy needs', in Govindjee (ed.), *Photosynthesis*, vol. 2 (New York: Academic Press, 1982), pp. 459–95. (Govindjee does not use a first name.)

6 With thanks to Mike Raupach and Roger Gifford for stressing this point.

7 The main source for the information in this section is a six-part series of articles by Hugh Powell on fertilizing the ocean with iron, in *Oceanus*, Online Magazine of Research from Woods Hole Oceanographic Institution (2007), at http://www.whoi.edu/oceanus/viewArticle.do?id=34167 (accessed Feb. 2012).

8 P. W. Boyd et al., 'Mesoscale iron enrichment experiments 1993–2005: Synthesis and future directions', *Science*, 315 (2 Feb. 2007), pp. 612–17.

9 Hugh Powell, 'Will ocean iron fertilization work?', *Oceanus*, 7 Jan. 2008, at http://www.whoi.edu/oceanus/viewArticle.do?id=35609&sectionid=1000 (accessed Aug. 2012).

10 Robert Anderson quoted in Hugh Powell, 'Fertilizing the ocean with iron', *Oceanus*, 13 Nov. 2007, at http://www.whoi.edu/oceanus/viewArticle.do?id=34167 (accessed Feb. 2012).

11 Wendy Zukerman, 'Whale poop is vital to ocean's carbon cycle', *New Scientist*, 22 Apr. 2010.

12 Michael A. Borowitzka, 'Options for enhancement of oceanic CO_2 uptake', presentation to a conference organized by the Australian Academy of Science, Canberra, 26 Sept. 2011.

13 Or, more accurately, a major component of the smell of the sea; see http://www.uea.ac.uk/env/marinegas/research/dmsmain.shtml (accessed Jan. 2012).

14 See http://www.antarctica.gov.au/about-us/publications/australian-antarctic-magazine/2001–2005/issue-4-spring-2002/what-is-the-southern-ocean (accessed Aug. 2012).

15 Antarctic Climate and Ecosystems Cooperative Research Centre, *Position Analysis: Climate Change and the Southern Ocean* (Hobart: ACE CRC, 2011), at http://www.acecrc.org.au/access/repository/resource/4f15b7ba-6abc-102f-a3d0-40404adc5e91/ACE_OCEANS_POSITION_ANALYSIS_LOW_RES.pdf (accessed Jan. 2012).

16 Ibid., p. 4.

17 D. W. R. Wallace et al., *Ocean Fertilization: A Scientific Summary for Policy Makers* (Paris: Intergovernmental Oceanographic Commission/UNESCO, 2010), figure 4, at http://www.us-ocb.org/publications/190674e.pdf (accessed Jan. 2012).

18 Ian Jones, 'Contrasting micro- and macro-nutrient nourishment of the ocean', *Marine Ecology Progress Series*, 425 (2011), p. 290.

19 T. Lenton and N. Vaughan, 'The radiative forcing potential of different climate geoengineering options', *Atmospheric Chemistry and Physics*, 9 (2009), pp. 5539–61,

count the fertilization effect of phosphorus run-off as a form of inadvertent geo-engineering because it stimulates algal growth. However, the carbon fixed in the algae must find a permanent home in the ocean if it is to work, and that is doubtful.

20 James J. Elser, 'A world awash with nitrogen', *Science,* 334:6062 (16 Dec. 2011), pp. 1504–5.

21 J. Rockström et al., 'Planetary boundaries: Exploring the safe operating space for humanity', *Ecology and Society,* 14:2 (2009), p. 32, at http://www.ecologyandsociety.org/vol14/iss2/art32/ (accessed Jan. 2012).

22 Jacob Koshy, 'Iron filings may not be a magic fix for carbon dioxide', Livemint.com, 10 Oct. 2011, at http://www.livemint.com/2011/10/09230723/Iron-filings-may-not-be-a-magi.html (accessed Jan. 2012).

23 Victor Smetacek et al., 'Deep carbon export from a Southern Ocean iron-fertilized diatom bloom', *Nature,* 487 (19 July 2012), pp. 313–19.

24 Rockström et al., 'Planetary boundaries'.

25 H. S. Kheshgi, 'Sequestering atmospheric carbon dioxide by increasing ocean alkalinity', *Energy,* 20 (1995), pp. 915–22.

26 Lisa Hanle, Kamala Jayaraman and Joshua Smith, 'CO_2 emissions profile of the US cement industry', at http://www.epa.gov/ttnchie1/conference/ei13/ghg/hanle.pdf (accessed Dec. 2011).

27 Supporters of geoengineering solutions argue that the more we cut emissions the higher will be the costs of abatement. So it will be cheaper to retire the first coal-fired power plants than the last ones because the first ones will be old clunkers. As we rise up the cost curve, they argue, at some point it may become cheaper to lime the oceans than phase out more coal-fired power plants. But when all of the other likely and potential effects of continued emissions plus ocean liming are taken into account, it's hard to see this argument ever working in practice.

28 See http://www.cquestrate.com/the-idea/detailed-description-of-the-idea (accessed Mar. 2012).

29 Ibid.

30 L. Harvey, 'Mitigating the atmospheric CO_2 increase and ocean acidification by adding limestone powder to upwelling regions', *Journal of Geophysical Research,* 113 (2008).

31 Ibid., p. 16.

32 Lenton and Vaughan, 'The radiative forcing potential of different climate geoengineering options', p. 2588.

33 One supporter notes that the oceans fall well short of their potential 'produc-tivity', so humans should aim to ramp it up, just as we have increased the productivity of land by adding chemicals like nitrogen fertilizers. Lime is routinely added to agricultural land to offset acidification due to the addition of nitrogen.

34 G. Rau, 'Geoengineering via chemical enhancement of ocean CO_2 uptake and storage or ignore ocean chemistry at our peril', response to National Academy of Sciences request for input on geoengineering concepts, Lawrence Livermore National Laboratory, Livermore, CA, June 2009.

35 Thanks to Roger Gifford for this point.

36 Greg Rau and Ken Caldeira, 'Enhanced carbonate dissolution: A means of sequestering waste CO_2 as ocean bicarbonate', *Energy Conversion & Management,* 40 (1999), pp. 1803–13.

37 Ibid., p. 1807.

38 Ibid., p. 1809.
39 Anil Ananthaswamy, 'Fix acid oceans by dumping alkali in them? Forget it', *New Scientist*, 16 Dec. 2011.
40 See the sources for table 1. Note that the figure for soil is an estimate for pre-industrial stocks, while the others are for 2010.
41 R. Gifford, 'Grassland carbon sequestration: Management, policy and economics', *Integrated Crop Management*, 11 (2010), pp. 33–56, at http://www.carbon-ranching.org/INTRO/Integrated_Crop_Management.pdf (accessed Apr. 2012).
42 Royal Society, *Geoengineering the Climate: Science, Governance and Uncertainty* (London: Royal Society, 2009), p. 12.
43 Ibid., pp. 11–12.
44 C. Greene, B. Monger and M. Huntley, 'Geoengineering: The inescapable truth of getting to 350', *Solutions Journal*, 1:5 (Oct. 2010).
45 Stuart White and Dana Cordell, 'Peak phosphorus: The sequel to peak oil', at http://phosphorusfutures.net/peak-phosphorus (accessed Jan. 2012).
46 F. Orr, 'Onshore geologic storage of CO_2', *Science*, 325 (25 Sept. 2009), pp. 1656–8.
47 See http://www.geology.sdsu.edu/how_volcanoes_work/Nyos.html (accessed Mar. 2009).
48 D. Schrag, 'Storage of carbon dioxide in offshore sediments', *Science*, 325 (25 Sept. 2009), pp. 1658–9.
49 Vaclav Smil quoted in Jeff Goodell, 'Coal's new technology: Panacea or risky gamble?', *Yale Environment 360* (Yale School of Forestry and Environmental Studies), 14 July 2008, at http://e360.yale.edu/feature/coals_new_technology_panacea_or_risky_gamble/2036/ (accessed Aug. 2012).
50 David Keith quoted in Marc Gunther, 'The business of cooling the planet', *CNN Money*, 7 Oct. 2011.
51 'Direct air capture of CO_2 with chemicals: A technology assessment for the APS panel on public affairs', American Physical Society, 1 June 2011.
52 Ibid.
53 Over that era it would extract from the air around 400 billion tonnes of carbon dioxide for which safe, permanent disposal sites would need to be found (using the new infrastructure that would need to be built to transport and inject it). It is estimated that in the Earth's crust there may be suitable geological formations to accommodate 200–2,000 billion tonnes of carbon dioxide. If carbon capture and storage technology proves effective then we would expect the most economical sites to be used up by coal-fired power plants before a network of air capture machines was built.
54 The quotation is from Harry Martinson's epic poem *Aniara* (Ashland: Story Line Press, 1999), p. 33.

Chapter 3 Regulating Sunlight

1 Antarctic Climate and Ecosystems, *Position Analysis*, p. 8.
2 Of incoming solar radiation, 30 per cent is reflected back into space by clouds and the Earth's surface. The Earth's temperature is highly sensitive to this fraction, so clouds are vital to climate dynamics. According to one leading expert, Veerabhadran Ramanathan: 'We have practically no theory of why the planet's albedo should be 0.30. Given this state of the field, and given the fact that clouds exert a large global cooling effect, we need a new approach to cut through the

current impasse on this fundamental problem in climate dynamics'; quoted in Bob Henson, 'Reflective research', *UCAR Quarterly* (University Corporation for Atmospheric Research) (Summer 2005).

3 Stephen Salter, personal communication 27 Feb. 2012; John Latham et al., 'Marine cloud brightening', paper submitted to *Philosophical Transactions of the Royal Society*, Nov. 2010, p. 16, at http://www.atmos.washington.edu/~robwood/papers/geoengineering/Latham-MCB-finalpaper.pdf (accessed Jan. 2012).

4 See John Latham at http://www.mmm.ucar.edu/people/latham/ (accessed Jan. 2012).

5 Mentioned in an earlier version of Latham et al., 'Marine cloud brightening'.

6 Royal Society, *Geoengineering the Climate*, p. 28.

7 Clive Cussler and Graham Brown, *The Storm* (London: Penguin, 2012).

8 Stephen Salter, 'Pseudo-random spray patterns for a world-wide transfer-function of cloud albedo control for the reversal of global warming', draft, Institute for Energy Systems, University of Edinburgh, 19 Apr. 2010.

9 According to Britain's Met Office: 'Seeding off the coast of South Africa leads to a knock-on effect which reduces Amazon rainforest rainfall by 30%. This could accelerate die-back of the forest, which is one of the world's major carbon stores, thus releasing huge amounts of carbon into the atmosphere.' See 'Geoengineering could damage Earth's ecosystems', 9 Sept. 2009, at http://www.metoffice.gov.uk/news/releases/archive/2009/geoengineering (accessed Jan. 2012). Other modelling shows a lower effect on Amazon precipitation; see Latham et al., 'Marine cloud brightening'.

10 Comment at symposium, The Atmospheric Science and Economics of Climate Engineering via Aerosol Injection, at Max Planck Institute for Chemistry, Mainz, Germany, 14–16 May 2012.

11 Latham et al., 'Marine cloud brightening', p. 10.

12 Philip Rasch, John Latham and Chih-Chieh (Jack) Chen, 'Geoengineering by cloud seeding: Influence on sea ice and climate system', *Environmental Research Letters*, 4 (Dec. 2009); Latham et al., 'Marine cloud brightening', p. 15.

13 Stephen Salter believes that interventions to warm the planet may prove useful 'if we ever need to prevent the next . . . ice age' (personal communication 27 Feb. 2012).

14 David Mitchell, personal communication 27 Feb. 2012.

15 'Excess water vapor near cirrus clouds puzzles scientists', *Science Daily*, 30 Nov. 2006, at http://www.sciencedaily.com/releases/2006/11/061130191409.htm (accessed Jan. 2012); K. N. Liou, 'Cirrus clouds and climate', paper, University of California at Los Angeles, n.d., at http://www.atmos.ucla.edu/~liougst/Cirrus_&_Climate.pdf (accessed Jan. 2012).

16 David Mitchell and William Finnegan, 'Modification of cirrus clouds to reduce global warming', *Environmental Research Letters*, 4 (2009).

17 David Mitchell, 'Manipulating cirrus cloud', presentation to symposium, The Atmospheric Science and Economics of Climate Engineering via Aerosol Injection, at Max Planck Institute for Chemistry, Mainz, Germany, 14–16 May 2012.

18 Please note that this suggestion is meant to be satirical.

19 T. Thordarson and S. Self, 'Atmospheric and environmental effects of the 1783–1784 Laki eruption: A review and reassessment', *Journal of Geophysical Research*, 108:D1 (2003).

20 Richard J. Payne, 'The "Meteorological Imaginations and Conjectures" of Benjamin Franklin', *North West Geography*, 10 (2010).

21 Alan Robock, Martin Bunzl, Ben Kravitz and Georgiy L. Stenchikov, 'A test for geoengineering?', *Science*, 327 (29 Jan. 2010).

22 Philip Rasch et al., 'An overview of geoengineering of climate using stratospheric sulphate aerosols', *Philosophical Transactions of the Royal Society A*, 366:1882 (13 Nov. 2008), p. 4013.

23 S. J. Smith et al., 'Anthropogenic sulfur dioxide emissions 1850–2005', *Atmospheric Chemistry and Physics*, 11 (2011), pp. 1101–16.

24 Clive Hamilton, *Requiem for a Species* (London: Earthscan, 2010), p. 181.

25 N. Mahowald, 'Aerosol indirect effect on biogeochemical cycles and climate', *Science*, 334 (11 Nov. 2011), p. 795.

26 Rasch et al., 'An overview of geoengineering of climate using stratospheric sulphate aerosols', p. 4012.

27 Ibid., p. 4015.

28 Scott Barrett, 'The incredible economics of geoengineering', *Environmental and Resource Economics*, 39 (2008), pp. 45–54. For a more cautionary approach see M. Goes, N. Tuana and K. Keller, 'The economics (or lack thereof) of aerosol geoengineering', *Climatic Change*, published online 5 Apr. 2011, at http://www3. geosc.psu.edu/~kzk10/Goes_cc_11.pdf (accessed Aug. 2012).

29 U. Niemeier, H. Schmidt and C. Timmreck, 'The dependency of geoengineered sulfate aerosol on the emission strategy', *Atmospheric Science Letters*, 12:2 (Apr. 2011), pp. 189–94.

30 Claudia Timmreck et al., 'Are volcanic eruptions a good analogy for climate engineering?', presentation to symposium, The Atmospheric Science and Economics of Climate Engineering via Aerosol Injection, at Max Planck Institute for Chemistry, Mainz, Germany, 14–16 May 2012.

31 Niemeier et al., 'The dependency of geoengineered sulfate aerosol on the emission strategy'.

32 Jeffrey Pierce et al., 'Efficient formation of stratospheric aerosol for climate engineering by emission of condensible vapor from aircraft', *Geophysical Research Letters*, 37:L18805 (2010).

33 K. Caldeira and L. Wood, 'Global and Arctic climate engineering: Numerical model studies', *Philosophical Transactions of the Royal Society A*, 366 (2008), pp. 4039–56.

34 Royal Society, *Geoengineering the Climate*, table 3.4.

35 H. Schmidt et al., 'Solar irradiance reduction to counteract radiative forcing from a quadrupling of CO_2: Climate responses simulated by four earth system models', *Earth System Dynamics*, 3 (2012), pp. 63–78.

36 Ibid., table 5.

37 Ibid., table 2.

38 Ibid., p. 73.

39 P. Heckendorn et al., 'The impact of geoengineering aerosols on stratospheric temperature and ozone', *Environmental Research Letters*, 4:4 (2009).

40 Ibid., p. 5. Schmidt et al. recognize this weakness in their approach: 'Solar irradiance reduction', p. 64.

41 Rasch et al., 'An overview of geoengineering of climate using stratospheric sulphate aerosols', p. 4015.

42 Schmidt et al., 'Solar irradiance reduction', p. 74.

43 M. Bollasina, Y. Ming and V. Ramaswamy, 'Anthropogenic aerosols and the weakening of the South Asian summer monsoon', *Science*, 334:6055 (28 Oct. 2011), pp. 502–5.

44 S. Tilmes, R. Müller and R. Salawitch, 'The sensitivity of polar ozone depletion to proposed geoengineering schemes', *Science*, 320 (30 May 2008), pp. 1201–4.
45 E. M. Volodin, S. V. Kostrykin and A. G. Ryaboshapko, 'Simulation of climate change induced by injection of sulfur compounds into the stratosphere', *Atmospheric and Oceanic Physics*, 47:4 (2011), pp. 430–8.
46 A. Ross and D. Matthews, 'Climate engineering and the risk of rapid climate change', *Environmental Research Letters*, 4 (Oct.–Dec. 2009).
47 Although the predicted fall-back may not occur if feedback effects, such as forest fires and methane emissions, have been triggered.
48 Reported in Ross and Matthews, 'Climate engineering and the risk of rapid climate change'.
49 Robock et al., 'A test for geoengineering?'.
50 Royal Society, *Geoengineering the Climate*, p. 36.
51 Rasch et al., 'An overview of geoengineering of climate using stratospheric sulphate aerosols', p. 4030.
52 David Keith, 'Photophoretic levitation of engineered aerosols for geoengineering', *Proceedings of the National Academy of Sciences*, online, 7 Sept. 2010.
53 Heckendorn et al., 'The impact of geoengineering aerosols', p. 2.
54 Caldeira and Wood, 'Global and Arctic climate engineering', p. 4045.
55 Ibid., table 3.
56 Eli Kintisch, *Hack the Planet* (Hoboken, NJ: John Wiley & Sons, 2010), p. 93.
57 Michael MacCracken, 'Potential applications of climate engineering technologies to moderation of critical climate change impacts', submission to Intergovernmental Panel on Climate Change Expert meeting on geoengineering, Lima, Peru, 2011, see http://science.org.au/natcoms/nc-ess/documents/GEsymposium.pdf (accessed Sept. 2012); Michael MacCracken, 'On the possible use of geoengineering to moderate specific climate change impacts', *Environmental Research Letters*, 4 (2009).
58 MacCracken, 'On the possible use of geoengineering', p. 10.
59 'Air quality and health', WHO factsheet, at http://www.who.int/mediacentre/factsheets/fs313/en/index.html (accessed Jan. 2012).
60 The oceans have absorbed another 0.3°C. See figures 1 and 2 in J. Hansen, M. Sato, P. Kharecha and K. Schuckmann, 'Earth's energy imbalance and implications', n.d., at http://www.columbia.edu/~jeh1/mailings/2011/20110415_EnergyImbalancePaper.pdf (accessed Jan. 2012).
61 The story is more complicated as black carbon from fossil fuel burning is mixed in with sulphur and organic aerosols. Black carbon absorbs solar radiation, while sulphate aerosols reflect it. So it is vital that black carbon is eliminated at the same time as sulphate aerosols. M. Ramana et al., 'Warming influenced by the ratio of black carbon to sulphate and the black-carbon source', *Nature Geoscience*, 3 (Aug. 2010).
62 MacCracken, 'On the possible use of geoengineering', p. 9.

Chapter 4 The Players and the Public

1 Including the landmark report of Royal Society in 2009, *Geoengineering the Climate*; its 2011 follow-up on the governance of solar radiation management, Royal Society, *Solar Radiation Management: The Governance of Research* (London: Royal Society, 2011); the report by the Washington think tank the Bipartisan Policy Center, Task Force on Climate Remediation Research, *Geoengineering: A National Strategic Plan for Research on the Potential Effectiveness, Feasibility, and*

Consequences of Climate Remediation Technologies (Washington, DC: Bipartisan Policy Center, 4 Oct. 2011); the report of the Novim Group, Jason Blackstock et al., *Climate Engineering Responses to Climate Emergencies* (Santa Barbara: Novim Group, 2009); Morgan and Ricke, 'Cooling the Earth through solar radiation management', for the International Risk Governance Council; the NASA workshop on managing solar radiation, Lee Lane, Ken Caldeira, Robert Chatfield and Stephanie Langhoff, 'Workshop report on managing solar radiation', Ames Research Center, NASA, California, Apr. 2007; and the UNESCO evaluation of geoengineering, see 'Geoengineering', at http://www.unesco.org/new/en/natural-sciences/environment/earth-sciences/emerging-issues/geo-engineering/ (accessed Aug. 2012). The UNESCO process incorporated a range of differing views. In its second process the Royal Society introduced some more sceptical voices (including mine).

2 Secretary of State for Energy and Climate Change, *Government Response to the House of Commons Science and Technology Committee Fifth Report of Session 2009–10: The Regulation of Geoengineering*, Cm 7936 (London: Stationery Office, Sept. 2010); Government Accountability Office, 'Climate engineering: Technical status, future directions, and potential responses', GAO, Washington, DC, 2011.

3 Kintisch, *Hack the Planet*. Over the period 1990–2010, nine scientists were responsible for 36 per cent of assertions made in the media about geoengineering, with Keith and Caldeira alone accounting for 15 per cent: Holly Jean Buck, 'What can geoengineering do for us? Public participation and the new media landscape', paper for workshop, The Ethics of Solar Radiation Management, 18 Oct. 2010, University of Montana, at http://www.umt.edu/ethics/EthicsGeoengineering/Workshop/articles1/Holly%20Buck.pdf (accessed Feb. 2012).

4 Jeff Goodell, *How to Cool the Planet* (Boston: Houghton Mifflin Harcourt, 2010), pp. 38–9. Goodell anticipates that Keith will be one of the 'superheroes of the geoengineering era' (p. 40).

5 Quoted in Jeff Goodell, 'Geoengineering: The prospect of manipulating the planet', *Yale Environment 360*, 7 Jan. 2009, at http://e360.yale.edu/feature/geoengineering_the_prospect_of_manipulating_the_planet/2107/ (accessed Aug. 2012).

6 See Henry Fountain, 'Trial balloon: A tiny geoengineering experiment', 17 July 2012, at http://green.blogs.nytimes.com/2012/07/17/trial-balloon-a-tiny-geoengineering-experiment/ (accessed July 2012).

7 Jeff Goodell, 'Can Dr. Evil save the world?', *Rolling Stone*, 15 Nov. 2006; Goodell, *How to Cool the Planet*, pp. 112–14, 118–29.

8 Joe Romm, 'Exclusive: Dysfunctional, lop-sided geoengineering panel tries to launch greenwashing euphemism, "Climate remediation"', 6 Oct. 2011, at http://thinkprogress.org/romm/2011/10/06/336676/geoengineering-panel-climate-remediation/ (accessed Jan. 2012).

9 Quoted in Catherine Brahic, 'Solar shield could be quick fix for global warming', *New Scientist*, 5 June 2007.

10 Joe Romm, 'Exclusive: Caldeira calls the vision of Lomborg's Climate Consensus "a dystopic world out of a science fiction story"', 5 Sept. 2009, at http://thinkprogress.org/climate/2009/09/05/204600/caldeira-delayer-lomborg-copenhagen-climate-consensus-geoengineering/ (accessed 23 July 2012).

11 Quoted in Brahic, 'Solar shield could be quick fix for global warming'.

12 See Goodell, *How to Cool the Planet*, p. 113 passim; Gunther, 'The business of cooling the planet'; Oliver Morton, 'Heroes of the environment 2009: David Keith', *Time*, 22 Sept. 2009, at http://www.time.com/time/specials/packages/article/0,28804,1924149_1924154_1924428,00.html (accessed Jan. 2012).

13 See 'Fund for Innovative Climate and Energy Research', at http://keith.seas.harvard.edu/FICER.html (accessed Feb. 2012)

14 Eli Kintisch, 'Bill Gates funding geoengineering research', *Science Insider*, 26 Jan. 2010, at http://news.sciencemag.org/scienceinsider/2010/01/bill-gates-fund.html (accessed Jan. 2012); 'Fund for Innovative Climate and Energy Research'.

15 Ben Webster, 'Bill Gates pays for "artificial" clouds to beat greenhouse gases', *Times Online*, 8 May 2010.

16 Latham et al., 'Marine cloud brightening', p. 16.

17 Gunther, 'The business of cooling the planet'.

18 Patent numbers US 2010/0034724 A1 and US 2010/0064890 A1, also WO/2010/022339.

19 Gunther, 'The business of cooling the planet'.

20 At http://www.intellectualventures.com/index.php/inventor-network (accessed Sept. 2012).

21 Goodell, *How to Cool the Planet*, p. 112.

22 Intellectual Ventures has been accused of 'patent hoarding' and is said to have lodged 20,000 patent applications; see 'Searete a part of Intellectual Ventures, a patent-trolling firm', 11 Dec. 2008, at http://techrights.org/2008/11/12/searete-under-intellectual-ventures/ (accessed Jan. 2012).

23 'The Stratospheric Shield' (2009), at http://intellectualventureslab.com/wp-content/uploads/2009/10/Stratoshield-white-paper-300dpi.pdf (accessed Aug. 2012).

24 The patent is owned by Searate LLC, which appears to be a subsidiary of Intellectual Ventures, see Todd Bishop, 'Bill Gates, top Microsoft executives do some of their inventing on the side', 10 Nov. 2008, at http://www.techflash.com/seattle/2008/11/Gates_top_Microsoft_executives_do_some_inventing_on_the_side34192179.html (accessed Jan. 2012). It has been accused of being a front company set up by Intellectual Ventures to hide the real owners, see 'Bill Gates' new career? Patent troll for Nathan Myhrvold?', at http://www.techdirt.com/articles/20081108/1744562771.shtml (accessed Feb. 2012). See also http://www.patentstorm.us/applications/20090173404/claims.html (accessed Jan. 2012). Rights for US patent 7655193 are assigned to Department of Energy and for US 20090173404 to an entity of the State of Delaware.

25 'Ken Caldeira on his Intellectual Ventures ties', 19 Oct. 2009, at http://warming101.blogspot.com.au/2009/10/ken-caldeira-on-his-intellectual.html (accessed Feb. 2012).

26 'The Stratospheric Shield', p. 14.

27 Joe Romm, 'Pro-geoengineering Bill Gates disses efficiency, "cute" solar, deployment – still doesn't know how he got rich', 5 May 2011, at http://thinkprogress.org/romm/2011/05/05/208032/bill-gates-efficiency-cute-solar/ (accessed Jan. 2012).

28 Joe Romm, 'Bill Gates disses energy efficiency, renewables, and near-term climate action while embracing the magical thinking of Bjorn Lomborg (and George Bush)', 26 Jan. 2010, at http://thinkprogress.org/romm/2010/01/26/205380/bill-gates-energy-efficiency-insulation-renewables-and-global-climate-action-bjorn-lomborg/ (accessed Jan. 2012).

29 Ken Caldeira, personal communication 29 June 2011.

30 'Virgin Earth Challenge', at http://www.virgin.com/subsites/virginearth/ (accessed Jan. 2012). His knighthood, he says, is 'fantastic proof' of his effectiveness, and Sir Richard has now set out on a quest to slay the dragon of climate change.

31 Helen Craig, 'Virgin Earth Challenge announces leading organisations', 2 Nov. 2011, at http://www.virgin.com/people-and-planet/blog/virgin-earth-challenge-announces-leading-organisations (accessed Jan. 2012).

32 See http://www.cquestrate.com/about-us (accessed Jan. 2012).

33 Blackstock et al., *Climate Engineering Responses to Climate Emergencies*, p. viii. On its influence see http://en.wikipedia.org/wiki/User:Tillman/Novim_Group (accessed Jan. 2012).

34 John Tierney, 'The Earth is warming? Adjust the thermostat', *New York Times*, 10 Aug. 2009, at http://www.nytimes.com/2009/08/11/science/11tier.html?_r=2 (accessed Feb. 2012).

35 On Exxon's campaign see James Hoggan, *Climate Cover-Up: The Crusade to Deny Global Warming* (Vancouver: Greystone Books, 2009), pp. 74–85. On the Royal Society's exasperation see David Adam, 'Royal Society tells Exxon: Stop funding climate change denial', *Guardian*, 20 Sept. 2006, at http://www.guardian.co.uk/environment/2006/sep/20/oilandpetrol.business (accessed Jan. 2012).

36 Jeffrey Ball, 'ExxonMobil thumbs its nose at climate change', *Wall Street Journal*, 15 June 2005. Also see http://www.heatisonline.org/contentserver/objecthandlers/index.cfm?id=5286&method=full (accessed Jan. 2012).

37 H. S. Kheshgi, 'Sequestering atmospheric carbon dioxide by increasing ocean alkalinity', *Energy*, 20 (1995), pp. 915–22.

38 'Climate change fears overblown, says ExxonMobil boss', *Guardian*, 28 June 2012, at http://www.guardian.co.uk/environment/2012/jun/28/exxonmobil-climate-change-rex-tillerson (accessed Aug. 2012).

39 The Kyoto Protocol (which the USA has not ratified) allows for emission reduction credits to be generated by abatement measures and the enhancement of carbon sinks, such as by reforestation.

40 The Planktos story is well told in both Goodell, *How to Cool the Planet*, pp. 148 passim, and Kintisch, *Hack the Planet*, ch. 7.

41 Goodell, *How to Cool the Planet*, p. 160.

42 See Shobita Parthasarathy et al., 'A public good? Geoengineering and intellectual property', Working Paper 10-1, Science, Technology, and Public Policy Program, Gerald R. Ford School of Public Policy, University of Michigan, June 2010, p. 6.

43 Hanna Wick, 'Protection from the Sun for planet Earth', *Neue Zürcher Zeitung*, 30 June 2012.

44 Parthasarathy et al., 'A public good?', p. 14.

45 Steven Shapin, 'Megaton man', *London Review of Books*, 24:8 (25 Apr. 2002).

46 'Arctic scientists warn of dangerous climate change', at http://insciences.org/article.php?article_id=10653 (accessed Jan. 2012).

47 Ian McInnes, 'The Arctic – gold rush tempered by harsh corporate realities', *Industrial Fuels and Power*, 27 Jan. 2012, formerly at http://www.ifandp.com/article/0015386.html (accessed Jan. 2012).

48 Rob Huebert, Heather Exner-Pirot, Adam Lajeunesse and Jay Gulledge, *Climate Change and International Security: The Arctic as a Bellwether* (Arlington, VA: Center for Climate and Energy Solutions, 2012).

49 Terry Macalister, 'Far north oil rush gambles with nature and diplomacy in game of high stakes', *Guardian*, 6 June 2012.

50 Huebert et al., *Climate Change and International Security*, p. 31.

51 MacCracken, 'Potential applications of climate engineering technologies'. However, it is worth noting that the call for immediate deployment of a solar filter over the Arctic by a group known as the Arctic Methane Emergency Group has been criticized as alarmist by other geoengineering researchers. See 'Climate scientists, geoengineering community slam AMEG', 19 Mar. 2012, at http://geoengineeringpolitics.blogspot.com.au/2012/03/climate-scientists-geoengineering.html (accessed Mar. 2012).

52 See http://www.sulphurinstitute.org/about/index.cfm (accessed Feb. 2012): 'The Sulphur Institute . . . is the global advocate for sulphur, representing all stakeholders engaged in producing, buying, selling, handling, transporting, or adding value to sulphur. We seek to . . . promote ongoing and uninterrupted efficient and safe handling and transport of all sulphur products while protecting the best interests of the environment.'

53 World production of sulphur in 2008 was 71.4 million tonnes, compared to an annual demand of perhaps 5 million tonnes for a programme of sulphate aerosol spraying. See 'Sulphur outlook', 12 Nov. 2008, at http://www.firt.org/sites/default/files/Clarke_Sulphur_Outlook_presentation.pdf (accessed Feb. 2012).

54 Work by Anthony Leiserowitz presented to the meeting at Asilomar in March 2010 and reported by Jeff Goodell, 'A hard look at the perils and potential of geoengineering', *Yale Environment 360*, 1 Apr. 2010, at http://e360.yale.edu/feature/a_hard_look_at_the_perils_and_potential_of_geoengineering/2260/ (accessed Feb. 2012). A survey by the US Congress's Government Accountability Office came to similar conclusions. After having geoengineering technologies outlined to them, around two-thirds of the respondents said they had not heard or read anything about them before: Government Accountability Office, 'Climate engineering', p. 62.

55 Masa Sugiyama, 'Climate engineering research in Japan', presentation to IMPLICC Symposium, 14–16 May 2012, at Max Planck Institute for Chemistry. Sugiyama makes the point that the survey was conducted prior to the 3/11 tsunami.

56 These data seem to be contradicted by an internet survey commissioned by David Keith and others, in A. Mercer, D. Keith and J. Sharp, 'Public understanding of solar radiation management', *Environmental Research Letters*, 6 (2011), but that research is implausible and not reported here.

57 Karen Parkhill and Nick Pidgeon, 'Public engagement on geoengineering research', Understanding Risk Working Paper 11-01, Cardiff University, June 2011.

58 Natural Environment Research Council, 'Experiment Earth? Findings from a Public Dialogue on Geoengineering', NERC, 2010.

59 Sugiyama, 'Climate engineering research in Japan'.

60 See http://www.nestanet.org/cms/node/2715 (accessed Feb. 2012).

61 Karen Parkhill and Nick Pidgeon, 'Public engagement on geoengineering research', Understanding Risk Working paper 11-01, Cardiff University, June 2011.

62 Buck, 'What can geoengineering do for us?', p. 7.

63 Natural Environment Research Council, 'The SPICE project: A geoengineering feasibility study', press release, 14 Sept. 2011, at http://www.nerc.ac.uk/press/releases/2011/22-spice.asp (accessed Feb. 2012).

64 National Academy of Sciences, *Advancing the Science of Climate Change* (Washington, DC: National Academies Press, 2010), p. 1.
65 Edward Maibach, Connie Roser-Renouf and Anthony Leiserowitz, 'Global warming's "six Americas" 2009: An audience segmentation', Yale Project on Climate Change and George Mason University Center for Climate Change Communication, 2009.
66 The story has been well told by Peter Jacques, Riley E. Dunlap and Mark Freeman, 'The organisation of denial: Conservative think tanks and environmental scepticism', *Environmental Politics*, 17:3 (June 2008), and Aaron McCright and Riley Dunlap, 'Anti-reflexivity: The American conservative movement's success in undermining climate science and policy', *Theory, Culture & Society*, 27:2–3 (2010), pp. 100–33.
67 See, e.g., Dan Kahan, 'Fixing the communications failure', *Nature*, 463 (21 Jan. 2010), pp. 296–7.
68 Naomi Klein, 'Capitalism vs. the climate', *The Nation*, 28 Nov. 2011.
69 Andrew Grice, 'Bush to G8: "Goodbye from the world's biggest polluter"', *Independent*, 10 July 2008, at http://www.independent.co.uk/news/world/politics/bush-to-g8-goodbye-from-the-worlds-biggest-polluter-863911.html (accessed Jan. 2012).
70 Neela Banerjee, 'Mitt Romney worked to combat climate change as governor', *Los Angeles Times*, 13 June 2012, at http://articles.latimes.com/2012/jun/13/nation/la-na-romney-energy-20120613 (accessed Aug. 2012).
71 Joe Romm, 'Call Jon Huntsman "crazy"', 7 Dec. 2011, at http://thinkprogress.org/romm/2011/12/07/383306/call-jon-huntsman-crazy-flips-on-climate-change-f-in-geography/ (accessed Jan. 2012).
72 Paul Douglas, 'A message from a Republican meteorologist on climate change', *Neorenaissance*, 28 Mar. 2012, at http://www.shawnotto.com/neorenaissance/blog20120329.html (accessed Mar. 2012).
73 A. McCright and R. Dunlap, 'Cool dudes: The denial of climate change among conservative white males in the United States', *Global Environmental Change*, 21 (2011), pp. 1163–72.
74 Ibid., figure 1.
75 Ibid., p. 1165.
76 Hamilton, *Requiem for a Species*, pp. 184–5.
77 Samuel Thernstrom, 'What role for geoengineering?', *The American* (American Enterprise Institute), 2 Mar. 2010, at http://www.american.com/archive/2010/march/what-role-for-geoengineering (accessed Mar. 2012).
78 Dan M. Kahan et al., 'Geoengineering and the science communication environment: A cross-cultural experiment', Working Paper 92, Yale Law School Cultural Cognition Project, 2012.
79 Ibid., p. 3.
80 Ibid., p. 19. The authors misunderstand the problem of 'moral hazard' as one of individual risk-taking instead of one inherent in political structures. Even so, at the individual level their demonstration that conservatives are more willing to accept risky forms of behaviour (geoengineering technologies) than safer ones (mitigation) if the 'costs' to their worldview are reduced seems to support the moral hazard argument.
81 Leslie Kaufman and Kate Zernike, 'Activists fight green projects, seeing UN plot', *New York Times*, 3 Feb. 2012.

82 See Clive Hamilton, 'Bullying, lies and the rise of right-wing climate denial', 22–26 Feb. 2010, p. 21, at http://www.clivehamilton.net.au/cms/media/documents/articles/abc_denialism_series_complete.pdf (accessed July 2012).

83 See http://today.uci.edu/news/2012/03/nr_rowlandobit_120312.php (accessed Mar. 2012).

84 Martin Enserink, 'Scientists ask minister to disavow predecessor's book', *Science*, 328 (9 Apr. 2010), p. 151.

85 An extended version of the Einstein story can be found in Clive Hamilton, 'What history can teach us about climate denial', in S. Weintrobe (ed.), *Engaging with Climate Change: Psychoanalytic Perspectives* (London: Routledge, 2012).

86 Jeroen van Dongen, 'Reactionaries and Einstein's fame: "German Scientists for the Preservation of Pure Science", relativity, and the Bad Nauheim meeting', *Physics in Perspective*, 9 (2007), p. 213.

87 Quoted by Jeroen van Dongen, 'On Einstein's opponents, and other crackpots', *Studies in History and Philosophy of Modern Physics*, 41 (2010), pp. 78–80.

88 David Rowe, 'Einstein's allies and enemies: Debating relativity in Germany, 1916–1920', in Vincent Hendricks et al. (eds), *Interactions: Mathematics, Physics and Philosophy, 1860–1930* (Dordrecht: Springer, 2006), p. 234.

89 Aaron McCright and Riley Dunlap, 'Defeating Kyoto: The conservative movement's impact on US climate change policy', *Social Problems*, 50:3 (2003), pp. 348–73; Jacques et al., 'The organisation of denial'.

90 Chris Mooney, *The Republican War on Science* (New York: Basic Books, 2005).

91 According to one conception, which had wide currency at the time, even among 'pro-Semites', Jews were 'innately inclined towards algorithmic, analytic, or abstract thinking, whereas Aryans tend to think intuitively and synthetically'. David Rowe, "Jewish mathematics" at Göttingen in the era of Felix Klein', *Isis*, 77:3 (Sept. 1986), p. 424.

92 McCright and Dunlap, 'Anti-reflexivity'.

93 Adding: 'The crisis of ecology, the threat of atomic war, and the disruption of the patterns of human life by advanced technology have all eroded what was once a general trust in the *goodness* of science': Jacob Needleman, *A Sense of the Cosmos: The Encounter of Modern Science and Ancient Truth* (New York: Doubleday, 1975), p. 1.

94 'Geo-engineering seen as a practical, cost-effective global warming strategy', 1 Dec. 2007, at http://news.heartland.org/newspaper-article/2007/12/01/geo-engineering-seen-practical-cost-effective-global-warming-strategy (accessed Jan. 2012).

95 On the media see, for example, Tim Lambert's Deltoid blog, at http://science-blogs.com/deltoid/global_warming/leakegate/ (accessed on Jan. 2012). On disgruntled scientists see Myanna Lahsen, 'Experiences of modernity in the greenhouse: A cultural analysis of a physicist "trio" supporting the backlash against global warming', *Global Environmental Change*, 18 (2008), pp. 204–19. On the role of the Koch brothers see Jane Mayer, 'Covert operations: The billionaire brothers who are waging a war against Obama', *New Yorker*, 30 Aug. 2010, at http://www.newyorker.com/reporting/2010/08/30/100830fa_fact_mayer (accessed Jan. 2012).

96 'Don't even mention global warming to kids', *Denver Post*, 25 May 2010.

97 'Oklahoma lawmaker Sally Kern proposes bill that forces teachers to question evolution', 28 Jan. 2011, at http://thinkprogress.org/politics/2011/01/28/140452/sally-kern-anti-evolution/ (accessed Jan. 2012).

98   Sara Reardon, 'Climate change sparks battles in classroom', *Science*, 333 (5 Aug. 2011), pp. 688–9. Also, Tennille Tracy, 'School standards wade into climate debate', *Wall Street Journal*, 11 Mar. 2012.

99   Reardon, 'Climate change sparks battles in classroom', p. 688.

100 Graham Readfearn, 'Plimer and Howard maintain the rage with climate science denial', 13 Dec. 2011, at http://www.readfearn.com/2011/12/plimer-and-howard-maintain-the-rage-with-climate-science-denial/ (accessed Jan. 2012).

101 Ian Enting, 'Ian Plimer's "Heaven + Earth": Checking the claims', at http://www.complex.org.au/tiki-download_file.php?fileId=91 (accessed Aug. 2012).

102 Adam Morton and Daniel Hurst, 'Libs want ban on teaching climate science', *Canberra Times*, 14 July 2012, at http://www.canberratimes.com.au/national/libs-want-ban-on-teaching-climate-science-20120713-221wj.html (accessed July 2012).

103 Clive Hamilton and Tim Kasser, 'Psychological adaptation to the threats and stresses of a four degree world', paper for Four Degrees and Beyond conference, University of Oxford, 28–30 Sept. 2009, at http://www.clivehamilton.net.au/cms/media/documents/articles/oxford_four_degrees_paper_final.pdf (accessed Jan. 2012).

104 Ian McEwan, *Solar* (London: Random House, 2010), p. 217.

105 Michael Spence, *The Next Convergence: The Future of Economic Growth in a Multispeed World* (New York: Farrar, Straus, & Giroux, 2011).

106 In his review of Spence's book, Harvard political economist Benjamin Friedman is equally oblivious: 'Whither China', *New York Review of Books*, 5 Apr. 2012.

107 Some of the following paragraphs draw on Hamilton, *Requiem for a Species*, pp. 118–33.

108 Renée Lertzman, 'The myth of apathy', *Ecologist*, 19 June 2008.

109 Kari Marie Norgaard, *Living in Denial: Climate Change, Emotions, and Everyday Life* (Cambridge, MA: MIT Press, 2010), p. 142.

110 Shelley Taylor, *Positive Illusions: Creative Self-Deception and the Healthy Mind* (New York: Basic Books, 1989).

111 Thanks are due to Rosemary Randall for stimulating these thoughts.

112 Winston Churchill, *Arms and the Covenant: Speeches by the Right Hon. Winston Churchill* (London: George C. Harrap, 1938), p. 297.

113 Roy Jenkins, *Churchill: A Biography* (New York: Farrar, Straus, & Giroux, 2001), p. 482.

114 Churchill, *Arms and the Covenant*, p. 171.

115 Ibid., pp. 152–3.

Chapter 5 Promethean Dreams

1 The story is well told by James Fleming, *Fixing the Sky: The Checkered History of Weather and Climate Control* (New York: Columbia University Press, 2010).

2 Ross Hoffman, 'Controlling the global weather', *Bulletin of the American Meteorological Society*, Feb. 2002, p. 241.

3 Ibid., p. 242.

4 Ibid., p. 246.

5 Douglas MacMynowski, 'Can we control El Niño?', *Environmental Research Letters*, 4 (2009).

6 Salter, 'Pseudo-random spray patterns'.

7 Ibid.

8 Rasch et al., 'Geoengineering by cloud seeding', p. 6.
9 Caldeira and Wood, 'Global and Arctic climate engineering', p. 4050.
10 Brad Allenby, 'Earth system engineering and management', *IEEE Technology and Society Magazine* (Institute of Electrical and Electronics Engineers) (Winter 2000/2001), pp. 10–24.
11 Ibid., p. 23 (emphasis added).
12 Ibid., pp. 22–3.
13 Brad Allenby, 'Geoengineering: A critique', *Potentials* (Institute of Electrical and Electronics Engineers), 31:1 (2012), pp. 22–6.
14 Prominent luke-warmists include Ted Nordhaus and Michael Shellenberger of the Breakthrough Institute, Roger Pielke Jr, Daniel Sarewitz, Steve Rayner, Mike Hulme and Bjorn Lomborg. See Clive Hamilton, 'Climate change and the soothing message of luke-warmism', *Climate Progress*, 25 July 2012, at http://thinkprogress.org/climate/2012/07/25/582051/climate-change-and-the-soothing-message-of-luke-warmism/ (accessed July 2012).
15 Blackstock et al., *Climate Engineering Responses to Climate Emergencies*, p. 19.
16 Ibid., p. 27.
17 Ibid., p. 28.
18 Ibid.
19 See, for example, David Spratt and Philip Sutton, *Climate Code Red* (Melbourne: Scribe, 2008).
20 Peaks like these can give a spurious sense of control since feedbacks from the biosphere (forest fires, methane emissions and so on) are wild cards that will probably drive concentrations beyond those directly arising from human activity.
21 Here I am drawing on Damon Matthews and Sarah Turner, 'Of mongooses and mitigation: Ecological analogues to geoengineering', *Environmental Research Letters*, 4 (2009).
22 Although the impacts of each case of biological control seem to be contested. See, for example, D. Simberloff and P. Stiling, 'Risks of species introduced for biological control', *Biological Conservation*, 78 (1996), pp. 185–92.
23 Quoted in Goodell, 'Can Dr. Evil save the world?'.
24 Quoted by Oliver Morton, 'Climate change: Is this what it takes to save the world?', *Nature*, 447 (10 May 2007), pp. 132–6. Ron Prinn is a professor at the Massachusetts Institute of Technology and a kiwi.
25 J. Eric Bickel and Lee Lane, 'An analysis of climate engineering as a response to climate change', Copenhagen Consensus Center, Copenhagen, 2009.
26 The paper is riddled with scientific errors, see Alan Robock, 'A biased economic analysis of geoengineering', 11 Aug. 2009, at http://www.realclimate.org/index.php/archives/2009/08/a-biased-economic-analysis-of-geoengineering/ (accessed Feb. 2012), all of which somehow conspire to support the authors' case. For example, Bickel and Lane defend their decision to ignore ocean acidification (which 'appears to be a potentially important matter') on the grounds that 'its relevance to CE [climate engineering] remains doubtful' ('An analysis of climate engineering as a response to climate change', p. 9). Yet their analysis of scenarios involving solar radiation management explicitly includes varying concentrations of carbon dioxide in the atmosphere, which must affect ocean acidification.
27 For a critique of the Nordhaus model see Hamilton, *Requiem for a Species*, pp. 56–62, and Clive Hamilton, 'Nordhaus's carbon tax: An excuse to do

nothing?' (2009), at http://www.clivehamilton.net.au/cms/media/critique_
of_nordhaus.pdf (accessed Aug. 2012).
28 See http://www.sdg.com/about-sdg/senior-staff/bickel (accessed Jan. 2012).
29 Bickel and Lane, 'An analysis of climate engineering as a response to climate
change', p. 8.
30 Lee Lane and David Montgomery, 'Political institutions and greenhouse gas
controls', American Enterprise Institute, Nov. 2008, p. 26.
31 They also acknowledge the assistance of Lowell Wood and Ken Caldeira.
32 Bickel and Lane, 'An analysis of climate engineering as a response to climate
change', p. 26.
33 Ibid., p. 11.
34 Buck, 'What can geoengineering do for us?'.
35 Kintisch, Hack the Planet, p. 98.
36 Hugh Gusterson, Nuclear Rites: A Weapons Laboratory at the End of the Cold War
(Berkeley: University of California Press, 1996).
37 Goodell, 'Can Dr. Evil save the world?'.
38 Peter Galison and Barton Bernstein, 'In any light: Scientists and the decision to
build the superbomb, 1952–1954', Historical Studies in the Physical and Biological
Sciences, 19:2 (1989), pp. 267–347.
39 William Broad, Teller's War (New York: Simon & Schuster, 1992), p. 20.
40 Ibid., p. 21.
41 Steven Shapin, 'Megaton man', London Review of Books, 24:8 (25 Apr. 2002).
Shapin wrote: 'What Teller liked, and what he was outstandingly good at, was
seeing the overall shape of very complex physical problems and making a series
of often totally wrong, occasionally brilliant guesses at how such problems might
be solved.'
42 Thanks are due to Henry Shue for reminding me of this fact.
43 Gusterson, Nuclear Rites, pp. 188–9.
44 Broad, Teller's War, p. 20.
45 Gusterson, Nuclear Rites, p. 204.
46 Ibid., p. 40.
47 Goodell, How to Cool the Planet, pp. 30–1, 126.
48 Gusterson, Nuclear Rites, p. 49.
49 Ibid., pp. 152 passim, 234.
50 Ibid., pp. 161–2.
51 Ibid., pp. 204, 197.
52 Goodell, How to Cool the Planet, p. 118 passim.
53 Gusterson, Nuclear Rites, p. 30.
54 Mikhail Gorbachev, 'Is the world really safer without the Soviet Union?', The
Nation, 9/16 Jan. 2012.
55 Goodell, How to Cool the Planet, p. 119.
56 E. Teller, L. Wood and R. Hyde, 'Global warming and ice ages: Prospects for
physics-based modulation of global change', Lawrence Livermore National
Laboratory, 15 Aug. 1997, p. 3.
57 Edward Teller, 'Sunscreen for Planet Earth', Wall Street Journal, 17 Oct. 1997.
58 E. Teller, R. Hyde and L. Wood, 'Active climate stabilization: Practical physics-
based approaches to prevention of climate change', Lawrence Livermore National
Laboratory, 18 Apr. 2002.
59 Ibid., p. 6.

60 The Hoover Institution promoted Thomas Gale Moore's *Climate of Fear: Why We Shouldn't Worry about Global Warming* (Washington, DC: Cato Institute, 1998). The Hoover Institution also promotes the work of virulently anti-green British columnist James Delingpole who has written that there 'aren't enough bullets' to kill all those who accept climate science: see http://blogs.telegraph.co.uk/news/jamesdelingpole/100069327/climate-change-there-just-arent-enough-bullets/ (accessed Feb. 2012). Among its 'distinguished visiting fellows' is Newt Gingrich.

61 Task Force on Climate Remediation Research, *Geoengineering*.

62 Robert Dreyfuss, 'Hawks, UAE ambassador want war with Iran', *The Nation*, 9 July 2010, at http://www.thenation.com/blog/37220/hawks-uae-ambassador-want-war-iran (accessed Feb. 2012).

63 Keith and Morgan were reported as opposing adoption of the term 'climate remediation'.

64 G. Prins et al., *The Hartwell Paper: A New Direction for Climate Policy after the Crash of 2009* (Oxford: Institute for Science, Innovation and Society, University of Oxford, 2010), at http://www2.lse.ac.uk/researchAndExpertise/units/mackinder/theHartwellPaper/Home.aspx. For a critique of the Hartwell report, see Hamilton, 'Climate change and the soothing message of luke-warmism'.

65 See http://www.bipartisanpolicy.org/projects/task-force-geoengineering/about (accessed Feb. 2012).

66 John Vidal, 'Big names behind US push for geoengineering', *Guardian* Environment Blog, 6 Oct. 2011, at http://www.guardian.co.uk/environment/blog/2011/oct/06/us-push-geoengineering (accessed Feb. 2012).

67 Romm, 'Exclusive'.

68 Eli Kintisch, 'DARPA to explore geoengineering', *Science Insider*, 14 Mar. 2009, at http://news.sciencemag.org/scienceinsider/2009/03/exclusive-milit.html (accessed Feb. 2012); Joe Romm, 'Memo to DARPA, Pentagon: Stay out of geoengineering–aka climate manipulation!', 16 Mar. 2009, at http://thinkprogress.org/romm/2009/03/16/203817/darpa-pentagon-military-geoengineering/ (accessed Feb. 2012).

69 Kintisch, 'DARPA to explore geoengineering'.

70 Robert Lempert and Don Prosnitz, 'Governing geoengineering research: A political and technical vulnerability analysis of potential near-term options', RAND Corporation, 2011.

71 Quoted by Goodell, *How to Cool the Planet*, p. 125.

72 James Fleming, 'The climate engineers', *Wilson Quarterly* (Spring 2007).

73 Fleming, *Fixing the Sky*, p. 166.

74 Tamzy J. House et al., 'Weather as a force multiplier: Owning the weather in 2025', research paper presented to Air Force 2025, Aug. 1996, at http://www.fas.org/spp/military/docops/usaf/2025/v3c15/v3c15-1.htm (accessed Sept. 2012).

75 Peter Schwartz and Doug Randall, 'An abrupt climate change scenario and its implications for United States national security', Oct. 2003, at http://www.climate.org/PDF/clim_change_scenario.pdf (accessed Mar. 2012).

76 K. Ricke, G. Morgan and M. Allen, 'Regional climate response to solar radiation management', *Nature Geoscience*, 3 (Aug. 2010), p. 537.

77 Fleming, 'The climate engineers'.

78 Myanna Lahsen, 'Experiences of modernity in the greenhouse: A cultural analysis of a physicist "trio" supporting the backlash against global warming', *Global Environmental Change*, 18 (2008), pp. 211–14.

79 Not only science but competition, property rights, medicine, consumerism and the work ethic define Western civilization: Niall Ferguson, *Civilization: The West and the Rest* (London: Penguin, 2011). For a devastating take-down see David Bromwich's review in 'The disappointed lover of the West', *New York Review of Books*, 8 Dec. 2011.

80 Lahsen, 'Experiences of modernity in the greenhouse', p. 216.

81 David Keith, 'Geoengineering the climate: History and prospect', *Annual Review of Energy and Environment*, 25 (2000), p. 279.

82 Ibid., p. 280.

Chapter 6 Atmospheric Geopolitics

1 Quoted in Keith, 'Geoengineering the climate', p. 252.

2 Quoted in ibid., p. 251.

3 Quoted by Goodell, *How to Cool the Planet*, p. 71.

4 Curiously, in the mid-1970s Teller was invited to Australia by mining magnate Lang Hancock to consult on how to create a deep-water harbour in Western Australia. Teller recommended a nuclear explosion (Tim Treadgold, 'Miner's daughter', *Forbes* magazine, 14 Feb. 2011). In 1992 Lang Hancock's mining company was inherited by his daughter, Gina Rinehart, who has recently become so rich that the business press are forecasting she will leapfrog Bill Gates and Carlos Slim to become the richest person in the world. Politically on the far right, Rinehart rejects climate science and sponsors trips to Australia by Christopher Monckton. In 2011 he delivered the Lang Hancock Memorial Lecture. She has recently bought media assets in order to exert political pressure. See 'Mining in a new vein', 2 Feb. 2012, at http://www.theage.com.au/opinion/politics/mining-in-a-new-vein-20120201-1qtcd.html (accessed Feb. 2012).

5 Y. Izrael et al., 'Field experiment on studying solar radiation passing through aerosol layers', *Russian Meteorology and Hydrology*, 34:5 (2009). The amateurishness of the reporting gives little confidence in the usefulness of the results.

6 'Why did IPCC's Yuri Izrael speak at the anti-IPCC conference in New York?', 4 Mar. 2008, at http://sandberghans.blogspot.com/2008/03/why-did-ipccs-yuri-izrael-speak-at-anti.html (accessed Nov. 2011).

7 Quirin Schiermeier and Bryon MacWilliams, 'Crunch time for Kyoto', *Nature*, 431 (2 Sept. 2004), pp. 12–13.

8 Ibid., p. 13.

9 Andrei Illarionov, 'Kyoto's smokescreen imperils us all', *Financial Times*, 14 Nov. 2004.

10 E. M. Volodin, S. V. Kostrykin and A. G. Ryaboshapko, 'Simulation of climate change induced by injection of sulfur compounds into the stratosphere', *Atmospheric and Oceanic Physics*, 47:4 (2011), pp. 430–8.

11 E. M. Volodin, S. V. Kostrykin, and A. G. Ryaboshapko, 'Climate response to aerosol injection at different stratospheric locations', *Atmospheric Science Letters*, 12:4 (2011), pp. 381–5.

12 Volodin et al., 'Simulation of climate change', p. 430.

13 Ibid., p. 438.

14 Kingsley Edney and Jonathan Symons, 'China and the blunt temptations of geoengineering: The role of solar radiation management in China's strategic response to climate change', *The Pacific Review*, forthcoming.

15 Daniela Yu and Anita Pugliese, 'Majority of Chinese prioritize environment over economy', 8 June 2012, at http://www.gallup.com/poll/155102/Majority-Chinese-Prioritize-Environment-Economy.aspx (accessed 18 June 2012).

16 Edney and Symons, 'China and the blunt temptations of geoengineering', p. 18.

17 Ibid., p. 36.

18 Ibid., p. 16.

19 Lennart Bengtsson, personal communication May 2012.

20 Andrew Parker, personal communication Nov. 2012.

21 Edney and Symons, 'China and the blunt temptations of geoengineering', p. 5.

22 Clive Hamilton, 'The end of the third world', e-International Relations, 29 Mar. 2010, at http://www.e-ir.info/2010/03/29/the-end-of-the-third-world/ (accessed Feb. 2012).

23 Edney and Symons, 'China and the blunt temptations of geoengineering', p. 27.

24 Joseph Needham, *Science and Civilisation in China*, vol. 4, part 3 (Cambridge: Cambridge University Press, 1971), pp. 234–5.

25 Tu Weiming, 'The continuity of being: Chinese visions of nature', in Mary Evelyn Tucker and John Berthrong (eds), *Confucianism and Ecology* (Cambridge, MA: Harvard University Press, 1998), pp. 105–21.

26 Asilomar Scientific Organizing Committee, *The Asilomar Conference Recommendations on Principles for Research into Climate Engineering Techniques* (Washington DC: Climate Institute, 2010), at http://climateresponsefund.org/images/Conference/finalfinalreport.pdf. Curiously, the sole 'strategic partner' (i.e. major funder) of the meeting was the government of the Australian state of Victoria, a state that relies almost solely on brown coal, the dirtiest kind, for electricity generation. The government planned to host a similar meeting in Australia but developed cold feet when it became aware of the optics of one of the worst polluters sponsoring a meeting on alternatives to emissions reductions. See Clive Hamilton, 'Victoria still talking to controversial geoengineering scientists', 10 June 2010, at http://www.crikey.com.au/2010/06/10/victoria-still-talking-to-controversial-geoengineering-scientists/ (accessed Feb. 2012).

27 Royal Society, *Geoengineering the Climate*, p. 40.

28 This idea is pushed especially by Granger Morgan; see for example Morgan and Ricke, 'Cooling the Earth through solar radiation management', for the International Risk Governance Council. The IRGC convened a meeting on geoengineering in Lisbon in April 2009 in collaboration with Morgan's Carnegie Mellon University and David Keith's University of Calgary.

29 See, for example, Richard Perkins, 'Technological "lock-in"', International Society for Ecological Economics and Internet Encyclopaedia of Ecological Economics, Feb. 2003, at http://www.ecoeco.org/pdf/techlkin.pdf (accessed Dec. 2011).

30 Secretary of State for Energy and Climate Change, *Government Response to the House of Commons Science and Technology Committee* (London: The Stationary Office, 2010).

31 Antarctic and Southern Ocean Coalition, '"No" to LOHAFEX's large-scale ocean fertilization', press release, 28 Jan. 2009, at http://www.scoop.co.nz/stories/WO0901/S00577.htm (accessed Feb. 2012).

32 Convention on Biological Diversity, 'Regulatory framework of climate-related geo-engineering relevant to the Convention on Biological Diversity', second draft, 23 Jan. 2012.

33 Although the ETC Group (Action Group on Erosion, Technology and Concentration) has proposed an International Convention for the Evaluation of New Technologies, a binding treaty that would assess, monitor and regulate new and emerging technologies based on a number of principles. See ETC Group, *Geopiracy: The Case against Geoengineering*, 2nd edn (Ottawa: ETC Group, Nov. 2010), appendix 2.

34 Thanks to Pat Mooney and Simon Terry for their observations from these meetings.

35 Stephen Leahy, 'Growing calls for moratorium on climate geoengineering', 26 Oct. 2010, at http://www.countercurrents.org/leahy261010.htm (accessed Feb. 2012).

36 Unpublished abstract of a paper delivered at the Peru meeting of the IPCC, June 2011.

37 Karen Scott, 'International law in the Anthropocene: Responding to the geoengineering challenge', *Michigan Journal of International Law*, forthcoming 2012.

38 William Burns, 'Climate geoengineering: Solar radiation management and its implications for intergenerational equity', *Stanford Journal of Law, Science and Policy*, published online May 2011, pp. 43–4.

39 See United Nations Office at Geneva, 'Confidence-Building Measures (CBMs)', at http://www.unog.ch/80256EE600585943/(httpPages)/CEC2E2D361ADFEE7 C12572BC0032F058? (accessed Feb. 2012).

40 See Joshua Horton, 'Geoengineering and the myth of unilateralism', *Stanford Journal of Law, Science and Policy*, published online May 2011. Horton takes the optimistic (but unpersuasive) view that unilateral action could not occur.

41 Lane et al., 'Workshop report on managing solar radiation'.

42 The Prometheans also included Tom Schelling and Tom Wigley. The few Soterians included James Fleming and Alan Robock.

43 This discussion draws on Lane et al., 'Workshop report on managing solar radiation,' pp. 10–13.

44 Blackstock et al., *Climate Engineering Responses to Climate Emergencies*, p. 1.

45 Lane et al., 'Workshop report on managing solar radiation', p. 11. The principal objective of the NASA meeting, like all meetings of geoengineers, was to secure more public funding for their research. The emergency framing, although politically attractive when the time came for deployment, might have made funding more difficult to secure. 'If deployment is perceived as lying many decades in the future, solar radiation management research projects might fare poorly in the contest for scarce research and development resources' (pp. 12, 13). Pre-emptive deployment offered a 'more direct rationale for near term research and development.' So they did not know which way to jump.

46 Ibid., p. 12. The same sentiments are found in Bickel and Lane, 'An analysis of climate engineering as a response to climate change', p. 26.

47 Lane et al., 'Workshop report on managing solar radiation', pp. 28–30.

48 Ibid., p. 29.

49 For example, in 2010 Ken Caldeira and David Keith wrote: 'Like it or not, a climate emergency is a possibility, and geoengineering could be the only affordable and fast-acting option to avoid a global catastrophe'. Caldeira and Keith, 'The need for climate engineering research', *Issues in Science and Technology* (Fall 2010). Keith has more recently backed away from the emergency argument.

Chapter 7 Ethical Anxieties

1 Crutzen, 'Albedo enhancement by stratospheric sulfur injections'.

2 Even so, all citizens of those nations with high per person greenhouse gas emissions must bear some responsibility for their nation's moral failure, even if they have taken all reasonable measures to reduce their own emissions. For a discussion see Steve Vanderheiden, *Atmospheric Justice: A Political Theory of Climate Change* (New York: Oxford University Press, 2008), pp. 173–80.

3 Royal Society, *Geoengineering the Climate*.

4 Crutzen, 'Albedo enhancement by stratospheric sulfur injections', p. 214.

5 See, e.g., Morgan and Ricke, 'Cooling the Earth through solar radiation management'; Caldeira and Keith, 'The need for climate engineering research'.

6 On this see Charles Taylor, 'Review of *The Fragility of Goodness* by Martha Nussbaum', *Canadian Journal of Philosophy*, 18:4 (Dec. 1988), p. 807.

7 For a discussion see Vanderheiden, *Atmospheric Justice*, p. 151.

8 Alan Robock, Luke Oman and Georgiy Stenchikov, 'Regional climate responses to geoengineering with tropical and Arctic SO_2 injections', *Journal of Geophysical Research*, 13 (2008).

9 Stephen Gardiner, 'Is "arming the future" with geoengineering really the lesser evil?', in Stephen Gardiner, Simon Caney, Dale Jamieson and Henry Shue (eds), *Climate Ethics: Essential Readings* (New York: Oxford University Press, 2010), p. 286.

10 For a discussion see Vanderheiden, *Atmospheric Justice*, p. 144 passim.

11 In their book *Climate Code Red*, David Spratt and Philip Sutton argue that sharply cutting emissions as soon as possible is essential but will not be enough. Geoengineering in the form of sulphate aerosol spraying may be needed to respond to an emergency and carbon dioxide removal methods will be needed to remove historical emissions.

12 Royal Society, *Geoengineering the Climate*, p. ix.

13 Rachel Carson, *Silent Spring* (1962; London: Penguin, 1965), p. 225.

14 See Clive Hamilton, 'The worldview informing the work of the Productivity Commission: A critique', 11 May 2006, at http://www.clivehamilton.net.au/cms/media/user_uploads/do_we_prefer_what_we_choose.pdf (accessed Feb. 2012).

15 Martin Weitzman, 'On modelling and interpreting the economics of catastrophic climate change', REStat (Renewable Energy Statistics Database) Final Version, 7 July 2008, p. 18.

16 In both the private and public domains, moral hazard involves taking insufficient care and engaging in more reckless behaviour. In the private domain, decision-makers are more reckless because the costs of their actions are spread onto others (it is therefore instrumentally rational). In the public domain, decision-makers are more reckless because they convince themselves that the risks are lower than they actually are (it is therefore irrational). The latter qualifies as moral corruption, the subversion of the moral discourse to one's own ends.

17 Naomi Oreskes and Erik Conway, *Merchants of Doubt* (London: Bloomsbury, 2010); Clive Hamilton, *Scorcher: The Dirty Politics of Climate Change* (Melbourne: Black Inc., 2007); Hoggan, *Climate Cover-Up*.

18 Royal Society, *Geoengineering the Climate*, pp. 39, 43.

19 Alex Steffen, 'Geoengineering and the new climate denialism', 29 Apr. 2009, at http://www.worldchanging.com/archives/009784.html (accessed Feb. 2012).

20 Steven D. Levitt and Stephen J. Dubner, *Superfreakonomics: Global Cooling, Patriotic Prostitutes, and Why Suicide Bombers Should Buy Life Insurance* (New York: HarperCollins, 2009). For a discussion of the legion of mistakes in the book begin here: http://www.guardian.co.uk/environment/2009/oct/19/superfreakonomics-geoengineering-wrong (accessed Feb. 2012).

21 At http://www.met.reading.ac.uk/Data/CurrentWeather/wcd/blog/at-the-controls-should-we-consider-geoengineering/ (accessed Feb. 2012).

22 Stephen Gardiner, 'Some early ethics of geoengineering the climate: A commentary on the values of the Royal Society report', *Environmental Ethics*, 20 (2011), pp. 163–88.

23 The next paragraphs draw from my *Requiem for a Species*, pp. 159–67.

24 See, for example, the World Coal Association on carbon capture and storage: 'Failure to widely deploy CCS will seriously hamper international efforts to address climate change', at http://www.worldcoal.org/carbon-capture-storage/ (accessed Feb. 2012).

25 'Trouble in store', *The Economist*, 5 Mar. 2009.

26 See http://www.youtube.com/watch?v=GehK7Q_QxPc (accessed Feb. 2012).

27 Roland Nelles, 'Germany plans boom in coal-fired power plants despite high emissions', *Der Spiegel Online*, 21 Mar. 2007.

28 Matthew Franklin, 'Obama supports Rudd on clean coal', *Australian*, 26 Mar. 2009.

29 Nicholas Stern, *The Economics of Climate Change* (Cambridge: Cambridge University Press, 2007), p. 251.

30 Jeffrey Sachs, 'Living with coal: Addressing climate change', speech to the Asia Society, New York, 1 June 2009.

31 Garnaut, *The Garnaut Climate Change Review*, p. 392.

32 International Energy Agency, *Technology Roadmap: Carbon Capture and Storage* (Paris: IEA, 2009).

33 'Trouble in store'; 'Carbon capture, storage projects funded', 18 May 2009, at http://www.themoneytimes.com/20090518/carbon-capture-storage-projects-funded-id-1068423.html (accessed Feb. 2012).

34 Christian Kerr, 'Carbon capture to save industry', *Australian*, 13 May 2009.

35 'Trouble in store'.

36 Vaclav Smil, 'Long-range energy forecasts are no more than fairy tales', *Nature*, 453:154 (8 May 2008).

37 'The illusion of clean coal', *The Economist*, 5 Mar. 2009.

38 In August 2011 Peter Cook, Australia's leading expert advocate for CCS, resigned as director of the flagship CO_2 Co-operative Research Centre, declaring that he was disappointed with progress and that the technology had reached a crisis point; see http://www.abc.net.au/lateline/content/2011/s3285864.htm (accessed Feb. 2012).

39 'Shell's Barendrecht carbon-capture project canceled', 5 Nov. 2010, at http://www.captureready.com/EN/Channels/News/showDetail.asp?objID=2030 (accessed Aug. 2012).

40 'Queensland delays ZeroGen project', *Carbon Capture Journal*, 20 Dec. 2010, at http://www.carboncapturejournal.com/displaynews.php?NewsID=707&PHPSESSID= d2hv0cjlmijbes2p17jmelllp6 (accessed Feb. 2012).

41 Joel Kirkland, 'Australia's desire for cleaner coal falls prey to high costs', 23 Dec. 2010, at http://www.nytimes.com/cwire/2010/12/23/23climatewire-australias-desire-for-cleaner-coal-falls-pre-84066.html?pagewanted=1 (accessed Feb. 2012).
42 Royal Society, *Geoengineering the Climate*, p. 61.
43 Carson, *Silent Spring*, p. 73.
44 See http://www.energymatters.com.au/index.php?main_page=news_article&article_id=400 (accessed Feb. 2012).
45 Hamilton, *Requiem for a Species*, chs 3–5.
46 S. Matthew Liao, Anders Sandberg and Rebecca Roache, 'Human engineering and climate change', *Ethics, Policy and the Environment*, special section, The Ethics of Geoengineering (July 2012), pp. 206–21. See also Leo Hickman, 'Bioengineer humans to tackle climate change, say philosophers', *Guardian* Environment Blog, 14 Mar. 2012; Ross Andersen, 'How engineering the human body could combat climate change', *The Atlantic*, 12 Mar. 2012.
47 Liao quoted in Andersen, 'How engineering the human body could combat climate change'.
48 Benjamin Hale quoted by Hickman, 'Bioengineer humans to tackle climate change, say philosophers'.
49 Gardiner, 'Some early ethics of geoengineering the climate', pp. 173–4.
50 The first argument is non-consequentialist and the second concerns consequences.
51 C. A. J. Coady, 'Playing God', in Julian Savulescu and Nick Bostrom (eds), *Human Enhancement* (Oxford: Oxford University Press, 2009), p. 163.
52 Michael Sandel, 'The case against perfection: What's wrong with designer children, bionic athletes, and genetic engineering', in Savulescu and Bostrom, *Human Enhancement*, p. 78.
53 Although Coady sets up the playing God argument more clearly than others, the arguments he considers against genetic engineering – the idea that choosing the future is beyond human limits, the sense that interfering in human nature is wrong, a worry about human 'autonomy' and a claim that human nature has some intrinsic value – remain somewhat vague and point only indirectly to god-like aspirations. For the atheist, the playing God argument can never be articulated with precision as there is always something ineffable about the spatial metaphysics on which it is based.
54 Natural Environment Research Council, 'Experiment Earth?'.

Chapter 8 This Goodly Frame

1 William Burroughs, *Climate Change in Prehistory* (Cambridge: Cambridge University Press, 2005), p. 13.
2 Ibid., p. 19.
3 Ibid., p. 37.
4 Ibid., figure 2.10, p. 58.
5 Rising sea levels washed away or inundated evidence of many coastal communities, expunging them from the archaeological record. They may also be the source of myths of the Flood that crop up independently in various cultures. At times, sea levels rose with a surge due, for example, to a sudden release of meltwater from lakes that had formed behind ice sheets, causing seas to wash inland by as much as 10 kilometres within a few months; ibid., p. 220.

6 Ibid., p. 53.

7 Ibid., p. 241.

8 Ibid., p. 73.

9 The International Commission on Stratigraphy, the scientific body that formally adjudicates on the division of the Earth's geological history into eons, eras, periods, epochs and ages, has established a working group to advise it on whether the Anthropocene meets the criteria for formal declaration as a new epoch to succeed the Holocene. The decision-making process is expected to take several years.

10 Will Steffen, Jacques Grinevald, Paul Crutzen and John McNeil, 'The Anthropocene: Conceptual and historical perspectives', *Philosophical Transactions of the Royal Society A*, 369 (2011), pp. 842–67.

11 Paul Crutzen and Eugene Stoermer, 'The Anthropocene', International Geosphere-Biosphere Programme Newsletter 41, 2000. See also Paul Crutzen, 'Geology of mankind', *Nature*, 415 (3 Jan. 2002), p. 23; Jan Janzalasiewicz, Mark Williams, Will Steffen and Paul Crutzen, 'The new world of the Anthropocene', *Environmental Science and Technology*, 44 (2010), pp. 2228–31.

12 Erle C. Ellis, 'Anthropogenic transformation of the terrestrial biosphere', *Philosophical Transactions of the Royal Society A,* 369 (2011), pp. 1025, 1027.

13 Wildlife Conservation Society, *State of the Wild 2006: A Global Portrait of Wildlife, Wildlands and Oceans* (Washington, DC: Island Press, 2005), p. 16.

14 Ellis, 'Anthropogenic transformation of the terrestrial biosphere', p. 1027.

15 Rockström et al., 'A safe operating space for humanity'.

16 Ibid.

17 Archer, *The Long Thaw*, p. 1.

18 Ibid., esp. ch. 12, and Curt Stager, *Deep Future: The Next 10,000 Years of Life on Earth* (New York: Thomas Dunne Books, 2011), esp. chs 1 and 2.

19 See Stager, *Deep Future*, pp. 21–4; Archer, *The Long Thaw*, pp. 73–5. Also J. Zachos et al., 'Trends, rhythms, and aberrations in global climate 65 Ma to present', *Science*, 292 (27 Apr. 2001), pp. 686–93, figure 1.

20 Archer, *The Long Thaw*, p. 151.

21 Ibid., p. 156.

22 For a review see B. McGuire, 'Potential for a hazardous geospheric response to projected future climate changes', *Philosophical Transactions of the Royal Society A*, 368 (2010), pp. 2317–45, and McGuire's fascinating subsequent book, *Waking the Giant* (Oxford: Oxford University Press, 2012).

23 McGuire, 'Potential for a hazardous geospheric response'.

24 'How the Japan earthquake shortened the earth day', at www.space.com (accessed Nov. 2011).

25 McGuire, *Waking the Giant*, p. 241.

26 Here I am relying on: Martin Rudwick, *Bursting the Limits of Time: The Reconstruction of Geohistory in the Age of Revolution* (Chicago: University of Chicago Press, 2005); James Moore, 'Geology and interpreters of Genesis in the nineteenth century', in David Lindberg and Ronald Numbers (eds), *God and Nature: Historical Essays on the Encounter between Christianity and Science* (Berkeley: University of California Press, 1986); Martin Rudwick, 'The shape and meaning of Earth history', in Lindberg and Numbers, *God and Nature*; and, Robert Muir Wood, *The Dark Side of the Earth* (London: George Allen & Unwin, 1985).

27 Quoted by Chris Decaen, 'Galileo Galilei, scriptural exegete, and the Church of Rome, advocate of science', lecture to Thomas Aquinas College, Santa Paula, CA, 2010.

28 Moore, 'Geology and interpreters of Genesis', p. 326. 'Scriptural geologists', on the other hand, weighed the facts according to whether they conformed to the Bible.

29 Rudwick, 'The shape and meaning of Earth history', p. 301.

30 Quoted by Rudwick, Bursting the Limits of Time, p. 2.

31 Rudwick, Bursting the Limits of Time, p. 286.

32 Quoted by Rudwick, 'The shape and meaning of Earth history', p. 313. One of the pioneers of modern geology, Lyell was nevertheless a pious man who for decades could not accept Darwin's theory and only reluctantly acknowledged its power towards the end of his life.

33 Even so, the structural elements of cosmological thinking persisted in new forms, not least in Hegel's inscription of the unfolding of 'Spirit' into human development. Rudwick notes that Edward Gibbon's The Decline and Fall of the Roman Empire, published over the period 1776–88, was perhaps the first history to integrate a political narrative with attention to detailed documentation (Bursting the Limits of Time, p. 182).

34 The break was not clean; before the links between natural events and human affairs were banished they were first scientized – see Fabien Locher and Jean-Baptiste Fressoz, 'Modernity's frail climate: A climate history of environmental reflexivity', Critical Inquiry, 38 (Spring 2012).

35 E. H. Carr, What Is History? (Harmondsworth: Penguin, 1964), p. 134.

36 Jacob Burckhardt, Reflections on History (1868; Indianapolis: Liberty Classics, 1979), p. 31.

37 Dipesh Chakrabarty, 'The climate of history: Four theses', Critical Inquiry, 35 (Winter 2009).

38 R. G. Collingwood, quoted by Chakrabarty, 'The climate of history', p. 4.

39 To echo a phrase from novelist Edward St Aubyn.

40 Stager, Deep Future, pp. 234–5. To be fair, these comments come in the epilogue to Stager's otherwise excellent book, an addendum of ill-considered 'personal reflections' that was, I'll wager, urged on him by his publisher.

41 Edward Teller, Lowell Wood and Roderick Hyde, 'Global warming and ice ages: Prospects for a physics-based modulation of global change', paper submitted to the 22nd International Seminar on Planetary Emergencies, Lawrence Livermore National Laboratory, 1997, p. 1.

42 W. Ruddiman, 'The anthropogenic greenhouse era began thousands of years ago', Climatic Change, 61 (2003), pp. 261–93.

43 Paul Crutzen and Will Steffen, 'How long have we been in the Anthropocene era? An editorial comment', Climatic Change, 61 (2003), p. 253.

44 Russell Powell et al., 'The ethics of geoengineering (working draft)', Oxford Uehiro Centre for Practical Ethics, 2010, p. 6; see http://www.practicalethics.ox.ac.uk/__data/assets/pdf_file/0013/21325/Ethics_of_Geoengineering_Working_Draft.pdf (accessed Aug. 2012).

45 Erle Ellis, 'The planet of no return', Breakthrough Journal, 2 (Fall 2011). Ellis's views, and similar ones, are published by the luke-warmist Breakthrough Institute.

46 Emma Morris, Rambunctious Garden: Saving Nature in a Post-wild World (New York: Bloomsbury, 2011), pp. 2–3.

47 Erle Ellis, 'Neither good nor bad', *New York Times*, 23 May 2011, at http://www.nytimes.com/roomfordebate/2011/05/19/the-age-of-anthropocene-should-we-worry/neither-good-nor-bad (accessed Mar. 2012).

48 Ronald Bailey, 'Better to be potent than not', *New York Times*, 23 May 2011, at http://www.nytimes.com/roomfordebate/2011/05/19/the-age-of-anthropocene-should-we-worry/better-to-be-potent-than-not (accessed Mar. 2012). Bailey quotes the declaration of 'environmental visionary' Steward Brand, 'We are as gods and might as well get good at it', a syntactically odd juxtaposition of the oratorical and the semi-literate.

49 Although we should not forget that most spend their working lives employed by mining and oil companies, and unconsciously absorb their worldview.

50 Stager, *Deep Future*, p. 229.

51 Ibid., p. 239.

52 Immanuel Kant, 'What is enlightenment?', in Lewis White Beck (ed.), *On History: Immanuel Kant* (Indianapolis: Bobbs-Merrill, 1963), p. 3.

53 Brad Johnson, 'Inhofe: God says global warming is a hoax', 9 Mar. 2012, at http://thinkprogress.org/green/2012/03/09/441515/inhofe-god-says-global-warming-is-a-hoax/ (accessed Mar. 2012).

54 Donella Meadows et al., *The Limits to Growth* (London: Earth Island, 1972), p. 184.

55 See Hamilton, *Requiem for a Species*, pp. 42–6.

56 Peter Sloterdijk, *Neither Sun nor Death* (Cambridge, MA: MIT Press, 2010), p. 237.

57 It would be easy to support this claim by pointing to the violent diatribes of conservative commentators in the Murdoch press and *Fox News*, but more telling examples can be drawn from the ranks of others who claim to be environmentalists of some kind. For example, T. Nordhaus and M. Shellenberger, *Break Through: From the Death of Environmentalism to the Politics of Possibility* (New York: Houghton Mifflin, 2007); Mark Lynas, *The God Species* (London: Fourth Estate, 2011); Andrew Charlton, *Man-Made World: Choosing between Progress and Planet*, Quarterly Essay 44 (Melbourne: Black Inc., 2012).

Index

Index

Index

SEMIOSIS

SUE BURKE spent many years working as a reporter and editor for a variety of newspapers and magazines. A Clarion workshop alumnus, Burke has published more than thirty short stories in addition to working extensively as a literary translator. She now lives in Chicago.

@SueBurkeSpain
mount-oregano.livejournal.com